Per Sense

Britannica

ENCYCLOPÆDIA BRITANNICA EDUCATIONAL CORPORATION

Mathematics in Context is a comprehensive middle grades curriculum. It was developed in collaboration with the Wisconsin Center for Education Research, School of Education, University of Wisconsin–Madison and the Freudenthal Institute at the University of Utrecht, The Netherlands, with the support of National Science Foundation Grant No. 9054928.

National Science Foundation

Opinions expressed are those of the authors
and not necessarily those of the Foundation

ISBN 0-7826-1494-9
1 2 3 4 5 6 7 8 9 10   99 98 97

The *Mathematics in Context* Development Team

Mathematics in Context is a comprehensive middle grades curriculum. The National Science Foundation funded the National Center for Research in Mathematical Sciences Education at the University of Wisconsin–Madison to develop and field-test the materials from 1991 through 1996. The Freudenthal Institute at the University of Utrecht in The Netherlands is the main subcontractor responsible for the development of the student and assessment materials.

The initial version of *Per Sense* was developed by Marja van den Heuvel-Panhuizen and Leen Streefland. It was adapted for use in American schools by James A. Middleton and Margaret R. Meyer.

National Center for Research in Mathematical Sciences Education Staff

Thomas A. Romberg
Director

Joan Daniels Pedro
Assistant to the Director

Gail Burrill
Coordinator
Field Test Materials

Margaret R. Meyer
Coordinator
Pilot Test Materials

Mary Ann Fix
Editorial Coordinator

Sherian Foster
Editorial Coordinator

James A. Middleton
Pilot Test Coordinator

Project Staff

Jonathan Brendefur
Laura J. Brinker
James Browne
Jack Burrill
Rose Byrd
Peter Christiansen
Barbara Clarke
Doug Clarke
Beth R. Cole

Fae Dremock
Jasmina Milinkovic
Margaret A. Pligge
Mary C. Shafer
Julia A. Shew
Aaron N. Simon
Marvin Smith
Stephanie Z. Smith
Mary S. Spence

Freudenthal Institute Staff

Jan de Lange
Director

Els Feijs
Coordinator

Martin van Reeuwijk
Coordinator

Project Staff

Mieke Abels
Nina Boswinkel
Frans van Galen
Koeno Gravemeijer
Marja van den Heuvel-Panhuizen
Jan Auke de Jong
Vincent Jonker
Ronald Keijzer

Martin Kindt
Jansie Niehaus
Nanda Querelle
Anton Roodhardt
Leen Streefland
Adri Treffers
Monica Wijers
Astrid de Wild

Acknowledgments

Several school districts used and evaluated one or more versions of the materials: Ames Community School District, Ames, Iowa; Parkway School District, Chesterfield, Missouri; Stoughton Area School District, Stoughton, Wisconsin; Madison Metropolitan School District, Madison, Wisconsin; Milwaukee Public Schools, Milwaukee, Wisconsin; and Dodgeville School District, Dodgeville, Wisconsin. Two sites were involved in staff development as well as formative evaluation of materials: Culver City, California, and Memphis, Tennessee. Two sites were developed through partnership with Encyclopædia Britannica Educational Corporation: Miami, Florida, and Puerto Rico. University Partnerships were developed with mathematics educators who worked with preservice teachers to familiarize them with the curriculum and to obtain their advice on the curriculum materials. The materials were also used at several other schools throughout the United States.

We at Encyclopædia Britannica Educational Corporation extend our thanks to all who had a part in making this program a success. Some of the participants instrumental in the program's development are as follows:

Allapattah Middle School
Miami, Florida
Nemtalla (Nikolai) Barakat

Ames Middle School
Ames, Iowa
Kathleen Coe
Judd Freeman
Gary W. Schnieder
Ronald H. Stromen
Lyn Terrill

Bellerive Elementary
Creve Coeur, Missouri
Judy Hetterscheidt
Donna Lohman
Gary Alan Nunn
Jakke Tchang

Brookline Public Schools
Brookline, Massachusetts
Rhonda K. Weinstein
Deborah Winkler

Cass Middle School
Milwaukee, Wisconsin
Tami Molenda
Kyle F. Witty

Central Middle School
Waukesha, Wisconsin
Nancy Reese

Craigmont Middle School
Memphis, Tennessee
Sharon G. Ritz
Mardest K. VanHooks

Crestwood Elementary
Madison, Wisconsin
Diane Hein
John Kalson

Culver City Middle School
Culver City, California
Marilyn Culbertson
Joel Evans
Joy Ellen Kitzmiller
Patricia R. O'Connor
Myrna Ann Perks, Ph.D.
David H. Sanchez
John Tobias
Kelley Wilcox

Cutler Ridge Middle School
Miami, Florida
Lorraine A. Valladares

Dodgeville Middle School
Dodgeville, Wisconsin
Jacqueline A. Kamps
Carol Wolf

Edwards Elementary
Ames, Iowa
Diana Schmidt

Fox Prairie Elementary
Stoughton, Wisconsin
Tony Hjelle

Grahamwood Elementary
Memphis, Tennessee
M. Lynn McGoff
Alberta Sullivan

Henry M. Flagler Elementary
Miami, Florida
Frances R. Harmon

Horning Middle School
Waukesha, Wisconsin
Connie J. Marose
Thomas F. Clark

Huegel Elementary
Madison, Wisconsin
Nancy Brill
Teri Hedges
Carol Murphy

Hutchison Middle School
Memphis, Tennessee
Maria M. Burke
Vicki Fisher
Nancy D. Robinson

Idlewild Elementary
Memphis, Tennessee
Linda Eller

Jefferson Elementary
Santa Ana, California
Lydia Romero-Cruz

Jefferson Middle School
Madison, Wisconsin
Jane A. Beebe
Catherine Buege
Linda Grimmer
John Grueneberg
Nancy Howard
Annette Porter
Stephen H. Sprague
Dan Takkunen
Michael J. Vena

Jesus Sanabria Cruz School
Yabucoa, Puerto Rico
Andreíta Santiago Serrano

John Muir Elementary School
Madison, Wisconsin
Julie D'Onofrio
Jane M. Allen-Jauch
Kent Wells

Kegonsa Elementary
Stoughton, Wisconsin
Mary Buchholz
Louisa Havlik
Joan Olsen
Dominic Weisse

Linwood Howe Elementary
Culver City, California
Sandra Checel
Ellen Thireos

Mitchell Elementary
Ames, Iowa
Henry Gray
Matt Ludwig

New School of Northern Virginia
Fairfax, Virginia
Denise Jones

Northwood Elementary
Ames, Iowa
Eleanor M. Thomas

Orchard Ridge Elementary
Madison, Wisconsin
Mary Paquette
Carrie Valentine

Parkway West Middle School
Chesterfield, Missouri
Elissa Aiken
Ann Brenner
Gail R. Smith

Ridgeway Elementary
Ridgeway, Wisconsin
Lois Powell
Florence M. Wasley

Roosevelt Elementary
Ames, Iowa
Linda A. Carver

Roosevelt Middle
Milwaukee, Wisconsin
Sandra Simmons

Ross Elementary
Creve Coeur, Missouri
Annette Isselhard
Sheldon B. Korklan
Victoria Linn
Kathy Stamer

St. Joseph's School
Dodgeville, Wisconsin
Rita Van Dyck
Sharon Wimer

St. Maarten Academy
St. Peters, St. Maarten, NA
Shareed Hussain

Sarah Scott Middle School
Milwaukee, Wisconsin
Kevin Haddon

Sawyer Elementary
Ames, Iowa
Karen Bush Hoiberg

Sennett Middle School
Madison, Wisconsin
Brenda Abitz
Lois Bell
Shawn M. Jacobs

Sholes Middle School
Milwaukee, Wisconsin
Chris Gardner
Ken Haddon

Stephens Elementary
Madison, Wisconsin
Katherine Hogan
Shirley M. Steinbach
Kathleen H. Vegter

Stoughton Middle School
Stoughton, Wisconsin
Sally Bertelson
Polly Goepfert
Jacqueline M. Harris
Penny Vodak

Toki Middle School
Madison, Wisconsin
Gail J. Anderson
Vicky Grice
Mary M. Ihlenfeldt
Steve Jernegan
Jim Leidel
Theresa Loehr
Maryann Stephenson
Barbara Takkunen
Carol Welsch

Trowbridge Elementary
Milwaukee, Wisconsin
Jacqueline A. Nowak

W. R. Thomas Middle School
Miami, Florida
Michael Paloger

Wooddale Elementary Middle School
Memphis, Tennessee
Velma Quinn Hodges
Jacqueline Marie Hunt

Yahara Elementary
Stoughton, Wisconsin
Mary Bennett
Kevin Wright

Site Coordinators

Mary L. Delagardelle—Ames Community Schools, Ames, Iowa

Dr. Hector Hirigoyen—Miami, Florida

Audrey Jackson—Parkway School District, Chesterfield, Missouri

Jorge M. López—Puerto Rico

Susan Militello—Memphis, Tennessee

Carol Pudlin—Culver City, California

Reviewers and Consultants

Michael N. Bleicher
Professor of Mathematics
University of Wisconsin–Madison
Madison, WI

Diane J. Briars
Mathematics Specialist
Pittsburgh Public Schools
Pittsburgh, PA

Donald Chambers
Director of Dissemination
University of Wisconsin–Madison
Madison, WI

Don W. Collins
Assistant Professor of Mathematics Education
Western Kentucky University
Bowling Green, KY

Joan Elder
Mathematics Consultant
Los Angeles Unified School District
Los Angeles, CA

Elizabeth Fennema
Professor of Curriculum and Instruction
University of Wisconsin-Madison
Madison, WI

Nancy N. Gates
University of Memphis
Memphis, TN

Jane Donnelly Gawronski
Superintendent
Escondido Union High School
Escondido, CA

M. Elizabeth Graue
Assistant Professor of Curriculum and Instruction
University of Wisconsin–Madison
Madison, WI

Jodean E. Grunow
Consultant
Wisconsin Department of Public Instruction
Madison, WI

John G. Harvey
Professor of Mathematics and Curriculum & Instruction
University of Wisconsin–Madison
Madison, WI

Simon Hellerstein
Professor of Mathematics
University of Wisconsin–Madison
Madison, WI

Elaine J. Hutchinson
Senior Lecturer
University of Wisconsin–Stevens Point
Stevens Point, WI

Richard A. Johnson
Professor of Statistics
University of Wisconsin–Madison
Madison, WI

James J. Kaput
Professor of Mathematics
University of Massachusetts–Dartmouth
Dartmouth, MA

Richard Lehrer
Professor of Educational Psychology
University of Wisconsin–Madison
Madison, WI

Richard Lesh
Professor of Mathematics
University of Massachusetts–Dartmouth
Dartmouth, MA

Mary M. Lindquist
Callaway Professor of Mathematics Education
Columbus College
Columbus, GA

Baudilio (Bob) Mora
Coordinator of Mathematics & Instructional Technology
Carrollton-Farmers Branch Independent School District
Carrollton, TX

Paul Trafton
Professor of Mathematics
University of Northern Iowa
Cedar Falls, IA

Norman L. Webb
Research Scientist
University of Wisconsin–Madison
Madison, WI

Paul H. Williams
Professor of Plant Pathology
University of Wisconsin–Madison
Madison, WI

Linda Dager Wilson
Assistant Professor
University of Delaware
Newark, DE

Robert L. Wilson
Professor of Mathematics
University of Wisconsin–Madison
Madison, WI

TABLE OF CONTENTS

BRITANNICA
Mathematics in Context

Dear Teacher,

Welcome! *Mathematics in Context* is designed to reflect the National Council of Teachers of Mathematics Standards for School Mathematics and to ground mathematical content in a variety of real-world contexts. Rather than relying on you to explain and demonstrate generalized definitions, rules, or algorithms, students investigate questions directly related to a particular context and construct mathematical understanding and meaning from that context.

The curriculum encompasses 10 units per grade level. This unit is designed to be the seventh, following *Figuring All the Angles,* but the unit also lends itself to independent use—to introduce students to experiences that will enrich their understanding of percents.

In addition to the Teacher Guide and Student Books, *Mathematics in Context* offers the following components that will inform and support your teaching:

- *Teacher Resource and Implementation Guide,* which provides an overview of the complete system, including program implementation, philosophy, and rationale

- *Number Tools,* which is a series of blackline masters that serve as review sheets or practice pages involving number issues and basic skills

- *News in Numbers,* which is a set of additional activities that can be inserted between or within other units; it includes a number of measurement problems that require estimation.

- *Teacher Preparation Videos,* which present comprehensive overviews of the units to help with lesson preparation

Thank you for choosing *Mathematics in Context.* We wish you success and inspiration!

Sincerely,

The Mathematics in Context Development Team

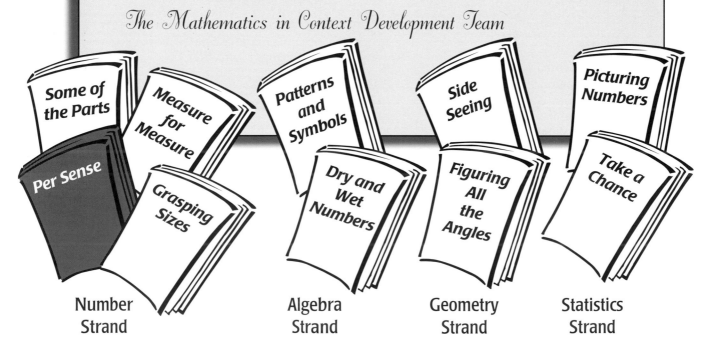

Number Strand   Algebra Strand   Geometry Strand   Statistics Strand

Overview

BRITANNICA

Mathematics in Context

36 km RACE

78%

78%

How to Use This Book

This unit is one of 40 for the middle grades. Each unit can be used independently; however, the 40 units are designed to make up a complete, connected curriculum (10 units per grade level). There is a Student Book and a Teacher Guide for each unit.

Each Teacher Guide comprises elements that assist the teacher in the presentation of concepts and in understanding the general direction of the unit and the program as a whole. Becoming familiar with this structure will make using the units easier.

Each Teacher Guide consists of six basic parts:

- Overview
- Student Material and Teaching Notes
- Assessment Activities and Solutions
- Glossary
- Blackline Masters
- Try This! Solutions

Overview

Before beginning this unit, read the Overview in order to understand the purpose of the unit and to develop strategies for facilitating instruction. The Overview provides helpful information about the unit's focus, pacing, goals, and assessment, as well as explanations about how the unit fits with the rest of the *Mathematics in Context* curriculum.

Note: After reading the Overview, view the Teacher Preparation Videotape that corresponds with the strand. The video models several activities from the strand.

Student Material and Teaching Notes

This Teacher Guide contains all of the student pages, each of which faces a page of solutions, samples of students' work, and hints and comments about how to facilitate instruction.

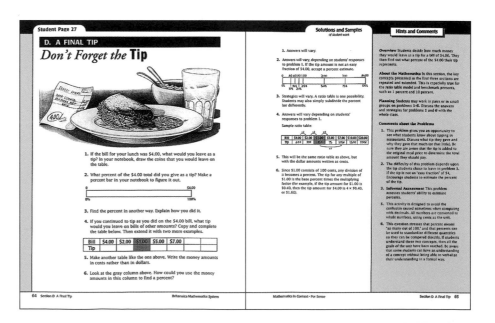

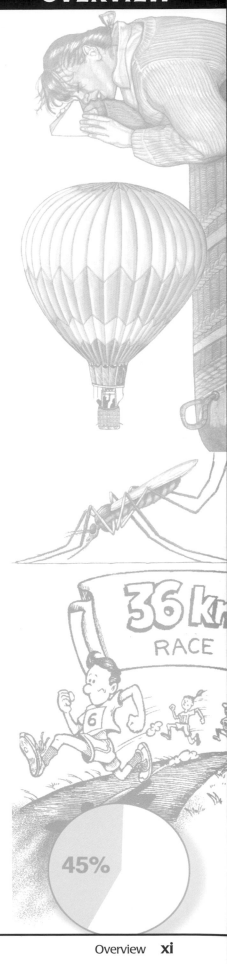

Each section within the unit begins with a two-page spread that describes the work students do, the goals of the section, new vocabulary, and materials needed, as well as providing information about the mathematics in the section and ideas for pacing, planning instruction, homework, and assessment.

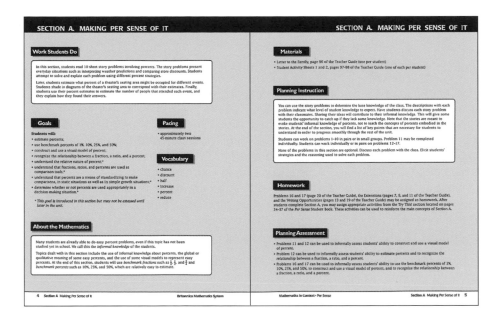

Assessment Activities and Solutions

Information about assessment can be found in several places in this Teacher Guide. General information about assessment is given in the Overview; informal assessment opportunities are identified on the teacher pages that face each student page; and the Assessment Activities section of this guide provides formal assessment opportunities.

Glossary

The Glossary defines all vocabulary words listed on the Section Opener pages. It includes mathematical terms that may be new to students, as well as words associated with the contexts introduced in the unit. (Note: The Student Book does not have a glossary. This allows students to construct their own definitions, based on their personal experiences with the unit activities.)

Blackline Masters

At the back of this Teacher Guide are blackline masters for photocopying. The blackline masters include a letter to families (to be sent home with students before beginning the unit), several student activity sheets, and assessment masters.

Try This! Solutions

Also included in the back of this Teacher Guide are the solutions to several Try This! activities—one related to each section of the unit—that can be used to reinforce the unit's main concepts. The Try This! activities are located in the back of the Student Book.

Unit Focus

Per Sense introduces students to experiences that will enrich their understanding of percents. The interrelationship between fractions, percents, and ratios is implicit throughout the unit. Students should develop an intuitive understanding of the concept of percent and its use as a standardization in order to make comparisons between different quantities. The major focus is on percents as a standardization of fractions and ratios in terms of parts per hundred. To build this understanding, students use percent bars and ratio tables.

The unit begins with 10 short stories that require students to make comparisons between different quantities. Using the context of a parking lot, the unit builds upon informal knowledge and continues with contexts involving baseball fans, tipping, and royalties on a book. These contexts develop mathematical models that students can use to estimate and calculate percents. The percent bar and ratio table are used to draw connections between representations of quantities and representations of percents. Algorithms are avoided as much as possible in this unit. Students will formalize their conceptions of quantities and operations more fully in later grades.

When students have finished *Per Sense,* they should recognize the need for standardization in order to make comparisons between different quantities. Students should be able to use percent bars, ratio tables, and benchmark percents, such as 1%, 10%, 25%, and 50%, to solve problems and to move between representations of quantities and representations of percents.

Mathematical Content

- recognizing percents as rational numbers or proportions
- relating percents to fractions and ratios
- using percents in situations of comparison and simple growth

Prior Knowledge

This unit assumes that students have:
- worked with fractions,
- seen and used the number line and ratio table,
- acquired an informal understanding and knowledge of percents,
- derived number knowledge from the unit *Some of the Parts.*

Facility with adding and subtracting three-digit numbers and with multiplying and dividing two-digit numbers is helpful for this unit. It is also helpful if students have had experience reading tables and using a ratio table or number line to perform computations.

Planning and Preparation

Pacing: 14 days

Section	Work Students Do	Pacing*	Materials
A. Making Per Sense of It	■ solve 10 short story problems using informal knowledge about percents	2 days	■ Letter to the Family (one per student) ■ Student Activity Sheets 1 and 2 (one of each per student)
B. Using Percents to Compare	■ compare the degree to which various parking lots are full using fractions, percent bars, ratio tables, and percents	4 days	■ Student Activity Sheets 3–8 (one of each per student) ■ meter stick (one per classroom)
C. Benchmark Percents	■ use percent bars to estimate percents for comparisons ■ develop different strategies for estimating percents ■ use percent benchmarks	4 days	■ Student Activity Sheets 9–12 (one of each per student)
D. A Final Tip	■ use ratio tables and benchmark percents such as 1% and 10%	4 days	

* One day is approximately equivalent to one 45-minute class session.

Preparation

In the *Teacher Resource and Implementation Guide* is an extensive description of the philosophy underlying both the content and the pedagogy of the *Mathematics in Context* curriculum. Suggestions for preparation are also given in the Hints and Comments columns of this Teacher Guide. You may want to consider the following:

- Work through the unit before teaching it. If possible, take on the role of the student and discuss your strategies with other teachers.

- Use the overhead projector for student demonstrations, particularly with overhead transparencies of the student activity sheets and any manipulatives used in the unit.

- Invite students to use drawings and examples to illustrate and clarify their answers.

- Allow students to work at different levels of sophistication. Some students may need concrete materials, while others can work at a more abstract level.

- Provide opportunities and support for students to share their strategies, which often differ. This allows students to take part in class discussions and introduces them to alternative ways to think about the mathematics in the unit.

- In some cases, it may be necessary to read the problems to students or to pair students to facilitate their understanding of the printed materials.

- A list of the materials needed for this unit is in the chart on page xiii.

- Try to follow the recommended pacing chart on page xiii. You can easily spend more time on this unit than the number of class periods indicated. Bear in mind, however, that many of the topics introduced in this unit will be revisited and covered more thoroughly in other *Mathematics in Context* units.

Resources

For Teachers	For Students
Books and Magazines *Mathematics Assessment: Myths, Models, Good Questions, and Practical Suggestions,* edited by Jean Kerr Stenmark (Reston, Virginia: The National Council of Teachers of Mathematics, Inc., 1991)	**Books and Magazines** *Number Tools,* Section H (component of *Mathematics in Context*)
Videos *Number Strand Teacher Preparation Video*	**Videos** Mathsphere Video *Jammin' in the USA* (available from Encyclopædia Britannica)

78%

Assessment

Planning Assessment

In keeping with the NCTM Assessment Standards, valid assessment should be based on evidence drawn from several sources. (See the full discussion of assessment philosophies in the *Teacher Resource and Implementation Guide*.) An assessment plan for this unit may draw from the following sources:

- Observations—look, listen, and record observable behavior.

- Interactive Responses—in a teacher-facilitated situation, note how students respond, clarify, revise, and extend their thinking.

- Products—look for the quality of thought evident in student projects, test answers, worksheet solutions, or writings.

These categories are not meant to be mutually exclusive. In fact, observation is a key part of assessing interactive responses and also key to understanding the end results of projects and writings.

Ongoing Assessment Opportunities

- **Problems within Sections**
  To evaluate ongoing progress, *Mathematics in Context* identifies informal assessment opportunities and the goals that these particular problems assess throughout the Teacher Guide. There are also indications as to what you might expect from your students.

- **Section Summary Questions**
  The summary questions at the end of each section are vehicles for informal assessment (see Teacher Guide pages 20, 38, 60, and 76).

End-of-Unit Assessment Opportunities

In the back of this Teacher Guide, there are seven assessments that can be completed in two 45-minute class periods. For a more detailed description of these assessment activities, see the Assessment Overview (Teacher Guide pages 78 and 79).

You may also wish to design your own culminating project or let students create one that will tell you what they consider important in the unit. For more assessment ideas, refer to the charts on pages xvi and xvii.

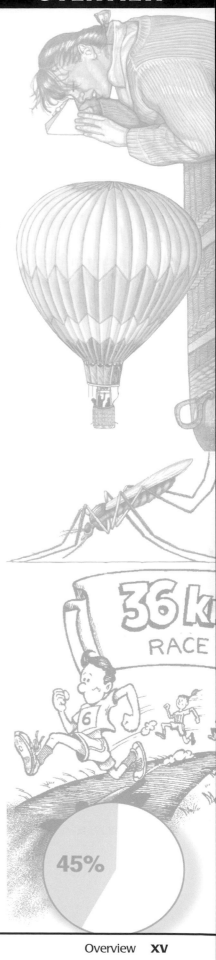

78%

Goals and Assessment

In the *Mathematics in Context* curriculum, unit goals, categorized according to cognitive procedures, relate to the strand goals and the NCTM Curriculum and Evaluation Standards. Additional information about these goals is found in the *Teacher Resource and Implementation Guide.* The *Mathematics in Context* curriculum is designed to develop students' abilities to perform with understanding in each of the categories. It is important to note that the attainment of goals in one category is not a prerequisite to attaining those in another category. In fact, students should progress simultaneously toward several goals in different categories.

	Goal	Ongoing Assessment Opportunities		End-of-Unit Assessment Opportunities
Conceptual and Procedural Knowledge	**1.** estimate percents	**Section A** **Section B** **Section C** **Section D**	p. 16, #12 p. 36, #20 p. 38, #22 p. 58, #24 p. 64, #3 p. 70, #12, #14	Keep It Clean, p. 80 Decorating the House, p. 86 On Loan, p. 88
	2. use benchmark percents of 1%, 10%, 25%, and 50%	**Section A** **Section B** **Section D**	p. 20, #16, #17 p. 36, #21 p. 70, #12, #14	Keep It Clean, p. 80 On Loan, p. 88 Parking Lots, p. 90
	3. construct and use a ratio table to find what percent is equivalent to a given fraction or ratio, or vice versa	**Section B** **Section C** **Section D**	p. 30, #14 p. 46, #8 p. 70, #12, #14	
	4. construct and use a visual model of percent	**Section A** **Section B** **Section C** **Section D**	p. 14, #11 p. 16, #12 p. 20, #16, #17 p. 36, #20 p. 38, #22 p. 46, #8 p. 70, #12, #14	Keep It Clean, ›. 80
	5. recognize the relationship between a fraction, a ratio, and a percent	**Section A** **Section B** **Section C** **Section D**	p. 16, #12 p. 20, #16, #17 p. 36, #20, #21 p. 38, #22 p. 46, #8 p. 70, #12, #14	Keep It Clean, p. 80

	Goal	Ongoing Assessment Opportunities		End-of-Unit Assessment Opportunities
Reasoning, Communicating, Thinking, and Making Connections	**6.** understand the relative nature of percent	**Section B** **Section C** **Section D**	p. 36, #21 p. 50, #12 p. 72, #15 p. 74, #19	The Best Buy, p. 82 Jammin', p. 84 Decorating the House, p. 86
	7. understand that fractions, ratios, and percents are used as comparison tools	**Section C** **Section D**	p. 50, #12 p. 58, #26 p. 74, #19	Jammin', p. 84 Parking Lots, p. 90
	8. understand that percents are a means of standardizing to make comparisons, in static situations as well as in simple growth situations	**Section A** **Section B**	p. 12, #8 p. 36, #20	Jammin', p. 84 Parking Lots, p. 90

	Goal	Ongoing Assessment Opportunities		End-of-Unit Assessment Opportunities
Modeling, Nonroutine Problem-Solving, Critically Analyzing, and Generalizing	**9.** determine what comparison tool is most appropriate in a given situation: fractions, ratios, or percents			Parking Lots, p. 90
	10. be able to determine which strategy for finding the percent is most appropriate in a given situation	**Section C**	p. 58, #24 p. 60, #27	Jammin', p. 84 Parking Lots, p. 90 Now It Is Your Turn, p. 92
	11. determine whether or not percents are used appropriately in a decision–making situation	**Section C** **Section D**	p. 50, #12 p. 58, #26 p. 74, #19	The Best Buy, p. 82

78%

More about Assessment

Scoring and Analyzing Assessment Responses

Students may respond to assessment questions with various levels of mathematical sophistication and elaboration. Each student's response should be considered for the mathematics that it shows, and not judged on whether or not it includes an expected response. Responses to some of the assessment questions may be viewed as either correct or incorrect, but many answers will need flexible judgment by the teacher. Descriptive judgments related to specific goals and partial credit often provide more helpful feedback than percentage scores.

Openly communicate your expectations to all students, and report achievement and progress for each student relative to those expectations. When scoring students' responses try to think about how they are progressing toward the goals of the unit and the strand.

Student Portfolios

Generally, a portfolio is a collection of student-selected pieces that is representative of a student's work. A portfolio may include evaluative comments by you or by the student. See the *Teacher Resource and Implementation Guide* for more ideas on portfolio focus and use.

A comprehensive discussion about the contents, management, and evaluation of portfolios can be found in *Mathematics Assessment: Myths, Models, Good Questions, and Practical Suggestions*, pp. 35–48.

Student Self-Evaluation

Self-evaluation encourages students to reflect on their progress in learning mathematical concepts, their developing abilities to use mathematics, and their dispositions toward mathematics. The following examples illustrate ways to incorporate student self-evaluations as one component of your assessment plan.

- Ask students to comment, in writing, on each piece they have chosen for their portfolios and on the progress they see in the pieces overall.

- Give a writing assignment entitled "What I Know Now about [a math concept] and What I Think about It." This will give you information about each student's disposition toward mathematics as well as his or her knowledge.

- Interview individuals or small groups to elicit what they have learned, what they think is important, and why.

Suggestions for self-inventories can be found in *Mathematics Assessment: Myths, Models, Good Questions, and Practical Suggestions*, pp. 55–58.

Summary Discussion

Discuss specific lessons and activities in the unit—what the student learned from them and what the activities have in common. This can be done in whole-class discussions, small groups, or in personal interviews.

Connections across the *Mathematics in Context* Curriculum

Per Sense is the third unit in the number strand. The map below shows the complete *Mathematics in Context* curriculum for grade 5/6. It shows where the unit fits in the number strand, and where it fits in the overall picture.

A detailed description of the units, the strands, and the connections in the *Mathematics in Context* curriculum can be found in the *Teacher Resource and Implementation Guide.*

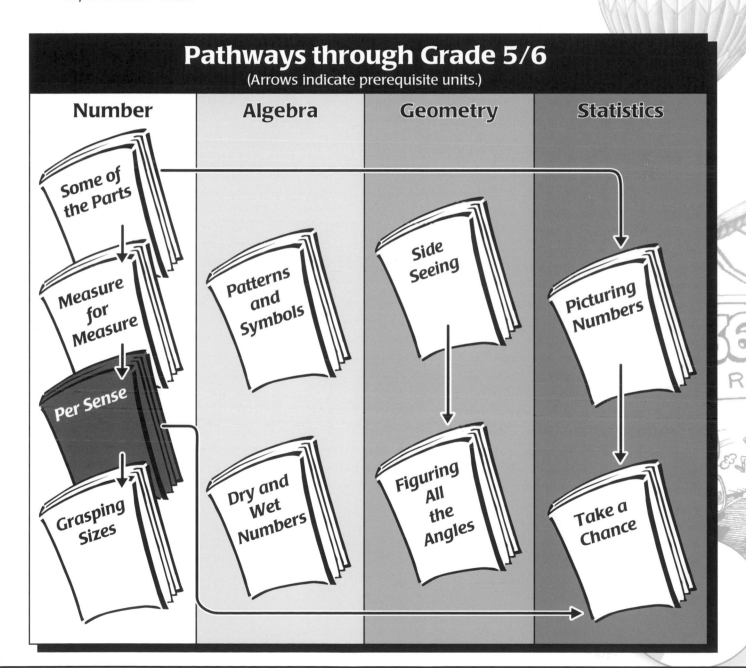

Pathways through Grade 5/6
(Arrows indicate prerequisite units.)

| Number | Algebra | Geometry | Statistics |

- Some of the Parts
- Measure for Measure
- Per Sense
- Grasping Sizes
- Patterns and Symbols
- Dry and Wet Numbers
- Side Seeing
- Figuring All the Angles
- Picturing Numbers
- Take a Chance

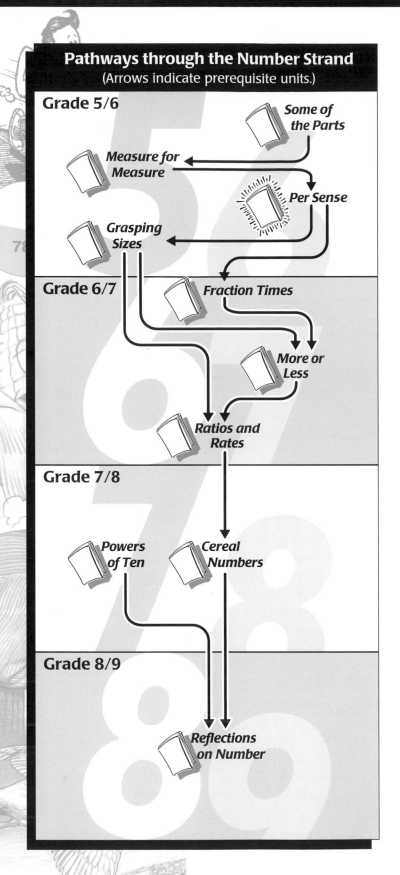

Pathways through the Number Strand
(Arrows indicate prerequisite units.)

Grade 5/6

Some of the Parts

Measure for Measure

Per Sense

Grasping Sizes

Grade 6/7

Fraction Times

More or Less

Ratios and Rates

Grade 7/8

Powers of Ten

Cereal Numbers

Grade 8/9

Reflections on Number

Connections within the Number Strand

On the left is a map of the number strand; this unit, *Per Sense,* is highlighted.

Per Sense is the third unit in the number strand and is preceded by *Some of the Parts* and *Measure for Measure.*

Some of the Parts extends the students' informal knowledge of part-whole relationships and provides opportunities for students to develop an informal understanding of operations with fractions. Ratio tables are introduced in this unit.

Measure for Measure allows students to develop an intuitive knowledge of decimals and provides a foundation for using the number line for estimation. Extra practice with the ratio table and number line can be found in the component called *Number Tools.*

Per Sense introduces students to the need for a standardization in order to make comparisons between different quantities. Expanded use of the percent bar and ratio table helps in drawing connections between representations of quantities and representations of percents.

Percents are revisited in *Fraction Times, More or Less,* and *Ratios and Rates.* More opportunities to use connections among fractions, decimals, and percents are found in the unit *Cereal Numbers.*

The Number Strand

Grade 5/6

Some of the Parts
Using fractions to describe the relative magnitude of quantities; ordering fractions; understanding and performing addition, subtraction, multiplication, and division operations with fractions.

Measure for Measure
Representing and using decimals in a variety of equivalent forms, investigating relationships among fractions and decimals, extending decimal number sense, adding and subtracting decimals.

Per Sense
Understanding percents as representing part-whole relationships; understanding the relationship between fractions, percents, and ratios; developing strategies for estimating and calculating percents.

Grasping Sizes
Developing a conceptual sense of ratio, estimating and calculating the effects of proportional enlargements or reductions, using scale lines, organizing data into ratio tables and calculating ratios, writing fractions as alternative expressions for equivalence situations.

Grade 6/7

Fraction Times
Comparing, adding, subtracting, and multiplying fractions; understanding the relationships among fractions, percents, decimals, and ratios.

More or Less
Connecting fractions, decimals, and percents; exploring percents as operators; discovering the effects of decimal multiplication.

Ratios and Rates
Relating ratio to fractions, decimals, and percents; dividing with decimals; differentiating between part-part and part-whole ratios; understanding the notions of rate, scale factor, and ratio as linear functions.

Grade 7/8

Cereal Numbers
Measuring volume and surface area in metric units; noting how changes in volume affect changes in the surface area of rectangular prisms; making comparisons with ratios, fractions, decimals, and percents; using a visual model to multiply with fractions; using a ratio strategy to divide with fractions.

Powers of Ten
Investigating simple laws for calculating with powers of ten, investigating very large and very small numbers.

Grade 8/9

Reflections on Number
Exploring primes, prime factorization, and divisibility rules; analyzing algorithms for multiplication and division; discovering and relating whole numbers, integers, and rational and irrational numbers by looking at the results of basic operations with their inverses.

Connections with Other *Mathematics in Context* Units

Percents are also an integral part of the statistics and probability strand, beginning informally with *Picturing Numbers,* in which students use 25%, 50%, and 100% to describe data. In *Take a Chance,* students express chance on a percent ladder. Students make decisions about experimental treatments using percent as a fundamental way to compare in *Great Expectations.* Simulations of population proportions, measured in percents, and describing data using percents occur in *Dealing with Data, Statistics and the Environment,* and *Insights into Data.*

The following mathematical topics that are included in the unit *Per Sense* are introduced or further developed in other *Mathematics in Context* units.

Prerequisite Topics

Topic	Unit	Grade
fractions, ratio table, fraction bar	*Some of the Parts*	5/6
number line, ratio table	*Number Tools*	5/6

Topics Revisited in Other Units

Topic	Unit	Grade
decimals and fractions	*Measure for Measure*	5/6
ratio table	*Grasping Sizes*	5/6
	Ratios and Rates	6/7
percent as comparison	*Take a Chance**	5/6
	Ratios and Rates	6/7
	Fraction Times	6/7
	Cereal Numbers	7/8
	*Great Expectations**	8/9
percent as operator	*Grasping Sizes*	5/6
	*Expressions and Formulas***	6/7
	Fraction Times	6/7
	Ratios and Rates	6/7
	*Growth***	8/9
relations between fractions, decimals, and percents	*Fraction Times*	6/7
	More or Less	6/7
	Ratios and Rates	6/7
	Cereal Numbers	7/8
percent as quantifier	*Grasping Sizes*	5/6
	*Dealing with Data**	6/7
	*Statistics and the Environment**	7/8
	*Looking at an Angle****	7/8
	*Insights into Data**	8/9

   * These units in the statistics and probability strand also help students make connections to ideas about numbers.

  ** These units in the algebra strand also help students make connections to ideas about numbers.

 *** These units in the geometry strand also help students make connections to ideas about numbers.

Student
Material
and Teaching
Notes

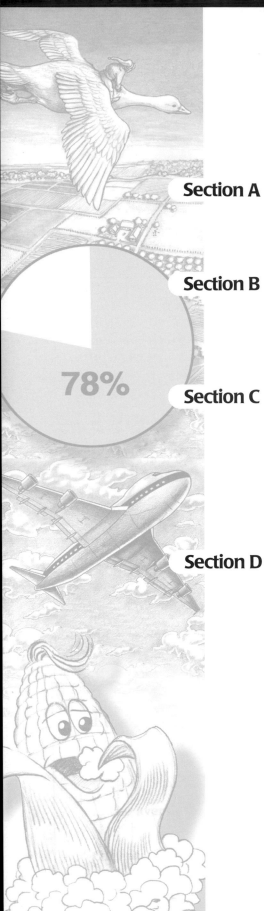

78%

Student Book
Table of Contents

Dear Student,

Welcome to *Per Sense*.

In this unit, you will use percents to compare things that are not easy to compare without percents, such as parking lots that have different numbers of total and occupied spaces.

You will estimate how many Dodgers fans are in this stadium using a percent.

You may even solve the mystery of *A Phony in Elbonia* with your knowledge of percents.

You will build and use tools called *percent bars* and *ratio tables* to help you find percents.

In the end, you should understand something about the way percents can help you compare different quantities and how to estimate percents.

Sincerely,

The Mathematics in Context Development Team

Work Students Do

In this section, students read 10 short story problems involving percents. The story problems present everyday situations such as interpreting weather predictions and comparing store discounts. Students attempt to solve and explain each problem using different percent strategies.

Later, students estimate what percent of a theater's seating area might be occupied for different events. Students shade in diagrams of the theater's seating area to correspond with their estimates. Finally, students use their percent estimates to estimate the number of people that attended each event, and they explain how they found their answers.

Goals

Students will:

- estimate percents;
- use benchmark percents of 1%, 10%, 25%, and 50%;
- construct and use a visual model of percent;
- recognize the relationship between a fraction, a ratio, and a percent;
- understand the relative nature of percent;*
- understand that fractions, ratios, and percents are used as comparison tools;*
- understand that percents are a means of standardizing to make comparisons, in static situations as well as in simple growth situations;*
- determine whether or not percents are used appropriately in a decision-making situation.*

  * This goal is introduced in this section but may not be assessed until later in the unit.

Pacing

- approximately two 45-minute class sessions

Vocabulary

- chance
- discount
- half
- increase
- percent
- reduce

About the Mathematics

Many students are already able to do easy percent problems, even if this topic has not been studied yet in school. We call this the *informal knowledge* of the students.

Topics dealt with in this section include the use of informal knowledge about percents, the global or qualitative meaning of some easy percents, and the use of some visual models to represent easy percents. At the end of this section, students will use *benchmark fractions* such as $\frac{1}{4}$, $\frac{1}{2}$, and $\frac{3}{4}$ and *benchmark percents* such as 10%, 25%, and 50%, which are relatively easy to estimate.

Materials

- Letter to the Family, page 96 of the Teacher Guide (one per student)
- Student Activity Sheets 1 and 2, pages 97–98 of the Teacher Guide (one of each per student)

Planning Instruction

You can use the story problems to determine the base knowledge of the class. The descriptions with each problem indicate what level of student knowledge to expect. Have students discuss each story problem with their classmates. Sharing their ideas will contribute to their informal knowledge. This will give some students the opportunity to catch up if they lack some knowledge. Note that the stories are meant to evoke students' informal knowledge of percents, not to teach the concepts of percents embodied in the stories. At the end of the section, you will find a list of key points that are necessary for students to understand in order to progress smoothly through the rest of the unit.

Students can work on problems 1–10 in pairs or in small groups. Problem 11 may be completed individually. Students can work individually or in pairs on problems 12–17.

None of the problems in this section are optional. Discuss each problem with the class. Elicit students' strategies and the reasoning used to solve each problem.

Homework

Problems 16 and 17 (page 20 of the Teacher Guide), the Extensions (pages 7, 9, and 11 of the Teacher Guide), and the Writing Opportunities (pages 13 and 19 of the Teacher Guide) may be assigned as homework. After students complete Section A, you may assign appropriate activities from the Try This! section located on pages 34–37 of the *Per Sense* Student Book. These activities can be used to reinforce the main concepts of Section A.

Planning Assessment

- Problems 11 and 12 can be used to informally assess students' ability to construct and use a visual model of percent.
- Problem 12 can be used to informally assess students' ability to estimate percents and to recognize the relationship between a fraction, a ratio, and a percent.
- Problems 16 and 17 can be used to informally assess students' ability to use the benchmark percents of 1%, 10%, 25%, and 50%, to construct and use a visual model of percent, and to recognize the relationship between a fraction, a ratio, and a percent.

A. MAKING PER SENSE OF IT

Percent *Situations*

In this section, you will see 10 short stories. Answer each question using a drawing or any other method to make your point clear.

Soccer or Band?

Joe's soccer team has practice on Wednesday afternoons. This week, the practice may be changed to Thursday.

"My goodness! What about your band practice on Thursday?" his mother asks.

"Don't worry. There is a 95 **percent chance** that soccer practice will still be on Wednesday," Joe replies.

1. Do you think this answer will put Joe's mother at ease? Why or why not?

Solutions and Samples
of student work

1. Yes, Joe's answer should put his mother at ease. A 95 percent chance is a near certainty that soccer practice will still be on Wednesday. Most students will realize that a 100 percent chance means certainty and that 95 percent is very close to 100 percent.

Hints and Comments

Overview Students discuss and explain 10 story problems that involve *percent.*

About the Mathematics Many students will be able to do easy problems on percents using their informal knowledge. Much of this informal knowledge depends on the exposure students have had to percents in daily life. These story problems are meant only to evoke students' informal knowledge of percents, not to teach the concepts of percents that are embodied in the stories.

Planning Students may work in pairs or in small groups on all 10 story problems. Let students discuss their ideas. Students' written explanations need not be organized in complete sentences. Observe where students are in their understanding of percent and use this knowledge as you continue in the unit. The Extension activity below may be assigned as homework.

Comments about the Problems

1. In this problem, students show their general knowledge about percent. Some students may draw a picture to express 95 percent. If so, the drawings do not need to be exact. It is sufficient that students understand that a 95 percent chance is a near certainty.

Extension Ask the following question:
If the weather forecast predicted a 95 percent chance of rain tomorrow, would you plan an outdoor picnic? [Answers may vary. Students might say "no," since this is almost a 100 percent chance of rain. Others may say "yes" due to the unpredictability of weather forecasts.] You do not need to confine the discussion to 95 percent. Other percents have their qualitative descriptions. For example, a 50 percent chance means that is as likely to rain as it is not to rain.

The Cross-Country Meet

Jill and Ann are talking about the cross-country meet next Saturday. The weather forecast predicts heavy rain during the meet. "I expect that **half** of the runners will give up before reaching the finish line," says Jill.

"I agree," says Ann. "I think about 60 percent will drop out."

2. Why does Ann say she agrees with Jill when Jill seems to be saying something different?

The Twin Discount

After twins Andre and Andrew had finished making their purchases at the music store, Andre said, "We both got a 10 percent **discount,** yet we didn't get the same dollar amount. I saved four dollars, and you saved only one dollar."

"Oh," said Andrew, "I can live with it."

3. Is it possible to get the same *percent* discount and not get the same *amount* of money? Explain your answer.

The Best Price Ever

A store advertised, "Best price ever! 40 percent discount on all items!"

Is it really the best price? The manager of the store says yes, but a customer says no.

4. Do you agree with the manager or the customer? Explain why.

2. Jill predicts that approximately one-half of the runners will drop out. Ann predicts approximately 60 percent of the runners will drop out. Since 60 percent is close to 50 percent (one-half), Ann is basically making the same prediction as Jill.

3. Yes, it is possible. A 10 percent discount on a higher-priced item represents a larger dollar amount saved than the dollar amount saved with a 10 percent discount on a lower-priced item. The only instance in which a 10 percent discount represents the same dollar amount saved is when both items have the same original price.

4. Answers will vary. Some students may agree with the manager, saying that it is very possible that a 40 percent discount allows for the best prices he has ever offered. Other students may agree with the customer, saying that the customer may have seen identical items on sale at lower prices at a different store. Some students may say that stores offering a 40 percent discount do not have the lowest prices since stores offering a 100 percent discount would have even better prices!

Overview Students continue to read and discuss story problems involving percent. In one story, students make a global comparison between one-half and 60 percent. They also encounter problems about percent and discount.

About the Mathematics The problems on this page deal with the following topics:

- The relationship between percents and fractions. For example, 50 percent is the same as one-half, and 25 percent is the same as one-quarter.

- Percents are relative numbers expressing a ratio. They are determined by the base against which they are calculated. For example, 50 percent of 20 (10) is not the same as 50 percent of 50 (25).

Planning Students may continue to work in pairs or in small groups on problems **2–4.** Have students discuss their ideas about each problem. Their explanations need not be given in neat, perfect sentences. Abbreviations, especially drawings, may be helpful.

Comments about the Problems

2. The general idea of this story is that percents can also be expressed as fractions. Avoid showing students how to convert percents to fractions (for example, 60 percent $= \frac{60}{100}$ or $\frac{6}{10}$) unless students themselves come up with them.

3. This problem deals with the relative nature of percent. Andre may have gotten a $40 pair of pants for $36, while Andrew may have purchased a $10 T-shirt for $9. Students do not have to find these amounts, but some may come up with them.

4. This story emphasizes the relative nature of percent. It is possible that a 20 percent discount will result in a better bargain than a 40 percent discount depending on the base price of the item and various factors such as markup and product quality.

Extension If students were able to understand problem **3,** you may ask them: *Would it work the other way around? Can two people get the same amount of discount and not the same percent discount?* [Yes. For example, suppose Andre buys an $8 tape. With a 50 percent discount, he saves $4. If Andrew buys a $10 tape with a 40 percent discount, he also saves $4.]

Budget Trouble

The student council president was explaining her budget plan: "This pie represents the school budget: 50 percent of the money goes for books, 25 percent for lunchroom improvements, and 35 percent for the student council president."

"Wait!" said the treasurer. "That adds up to 110 percent!"

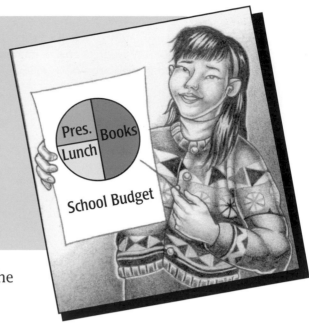

5. Is there a problem with the budget? Explain.

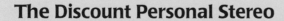

Saving Money

Last fall, Val started to save for a new bike. One day she counted her money and decided to keep on saving. A month later she counted her money again. When she finished, she shouted, "Wow, now I have over 200 percent!"

"Two hundred percent? That doesn't make sense," her brother said.

6. What could Val have meant by this? Give an example.

The Discount Personal Stereo

Liz: "Tom, look here. There is a 25 percent discount on this $80 personal stereo. You know, it's the one I wanted to buy. I need only $60 now."

Tom: "You need only $60? That doesn't seem right to me."

7. If you were Liz, how would you explain your answer to Tom?

5. Yes, there is a problem with the budget. The percents of the budget used for the three areas must add up to 100 percent. Some students might suggest allocating 50 percent of the budget for books and 25 percent each for the lunchroom improvements and the student council (not just the president).

6. Val could have meant that her present balance is twice the amount of her original balance (her total savings equals two times, or 200 percent, of her original balance).

7. Explanations will vary. One possibility follows:

The original price of the stereo is $80. It is on sale at a 25 percent discount. I know that 25 percent equals one-quarter. One-quarter of $80 is $20. So I subtract $20 from $80 to find the sale price of $60.

Overview Students encounter a situation in which more than 100 percent is not possible, followed by a situation that involves an increase of over 200 percent. In the third problem on this page, they make a simple computation with percents.

About the Mathematics The problems on this page focus on the following topics:

• the part-whole nature of percents, as in problem **5;**

• the visual representation of percents, as in a pie chart;

• positive and negative growth described by means of percent;

• percent used as an operator.

Planning Students can continue to work in pairs or in small groups on problems **5–7.** The Extension activity below may be assigned as homework.

Comments about the Problems

5. The key point implied in the story concerns the identification of 100 percent as the maximum. Find out how many students consider this concept to be self-evident. A budget over 100 percent only makes sense if you understand the concept of deficit spending.

6. Val's original account balance is considered to be the starting point, which is 100 percent. By saving twice as much as her original balance, Val has saved 200 percent.

7. Some students may use a different strategy here: determine the sale price as $\frac{3}{4}$ of the regular price.

Extension Have students solve this variation of problem **7.** *If Liz got a sale price of $40 instead of $60, what would be the percent of her discount?* [50 percent. Forty dollars is half of $80. Therefore, Liz is saving half, or 50 percent, of the original price.]

The Exam

In order to pass a test, Stephanie needed to answer 50% of the questions correctly. She was wrong on 14 problems.

8. Should we congratulate her or not? Explain.

The Increasing Rent

Dear Mom,

School is great, and I am doing well. I have some bad news! I can't believe it, but next month my rent will *increase* 25 percent! That means the end of my old $200 rent.

The manager hasn't told me what the exact rent will be, but I think I have to pay somewhere between two and three hundred dollars. Please send me $300 to be on the safe side ($100 extra this month).

Many kisses,
Juanita

9. How much should Juanita's mother send?

The Price War

Two shopkeepers are comparing their prices. Barbara's store sells a watch for $20. Dennis's store sells the same watch for $40. Barbara says, "Your store price is 100 percent more expensive!"

"That's not true," says Dennis. "Your store price is only 50 percent less."

10. Who is right?

8. It is impossible to know whether Stephanie should be congratulated or not. Since there is no mention of how many total problems were on the test, you cannot determine what fraction or percent of the total is represented by the 14 problems. Some students may make conjectures such as the following:

If there were 40 problems, half would be 20. Fourteen is fewer than 20, so we should congratulate her!

If there were 20 problems, half would be 10, and 14 is more than 10, so we should not congratulate her.

9. Answers will vary. Some students will calculate 25 percent of $200 to be $50 and say that Juanita's mother should send exactly $250. Other students may estimate 25 percent of $200 to be somewhere between $25–$100 and suggest that Juanita's mother send $225–$300.

10. Both shopkeepers are right. Barbara is correct in saying that Dennis's price of $40 is 100 percent higher than her price of $20. Dennis is also correct in saying that Barbara's price of $20 is 50 percent less than his price of $40. The two statements are equivalent.

Overview Students continue to read and discuss story problems involving percents. On this page, all of the problems focus on the relative nature of percent.

About the Mathematics Problems **8–10** are examples of the relative nature of percent. Percents are only understandable when compared to some base value.

Planning Students may continue to work in pairs or in small groups on problems **8–10.**

Comments about the Problems

8. If students are having difficulty, ask them the following questions:

   • *What if the exam consisted of 20 problems? Would Stephanie have succeeded then?* [No. $\frac{6}{20}$ = 30 percent]

   • *What if the exam had 70 problems?* [Yes. $\frac{56}{70}$ = 80 percent]

   Throughout the discussion, it may become more apparent to students that they need to know the total number of questions on the test. It is important that the students are aware of the fact that different levels of precision are necessary.

9. Allow students to compute or estimate the percent at the level of precision they can handle.

10. Ask students to view this situation from the perspective of each shopkeeper. The managers of the two stores are using different bases in calculating the percents they will use in advertising.

Writing Opportunity Use problem **9** as a writing activity. Invite students to assume the role of Juanita's mother, and have them write a letter to Juanita. The letter should include information about how much money Juanita's mother is sending her. The letter should also include the student's reasoning as to why he or she is sending that specific amount.

The School Theater

Stage	Audience

Malcolm Shabazz Middle School has a small theater. The stage has spotlights, curtains, and everything else you would expect to find in a large theater. The figure on the left shows the shape of the theater.

Some of the performances held there are more popular than others.

11. The following performances were given for the students. Use **Student Activity Sheet 1** to shade in the audience section of the theater to show what percent of the theater you think would be filled for each show.

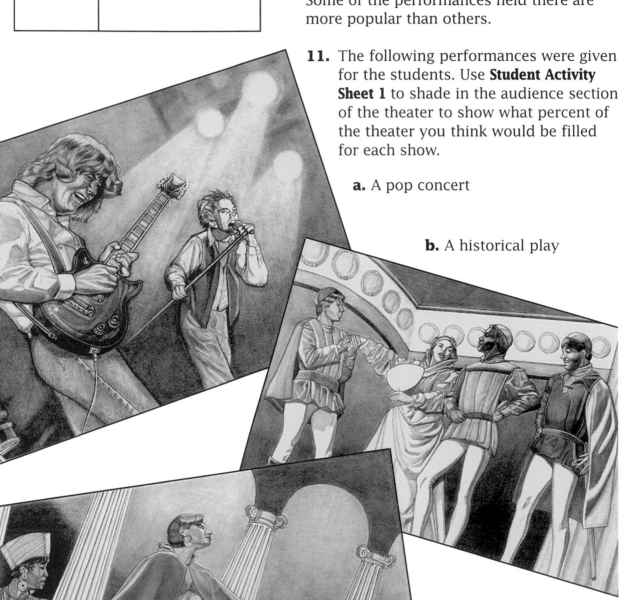

a. A pop concert

b. A historical play

c. A fashion show

11. Answers will vary. Accept all responses, as long as the shaded portion of the picture and the percent roughly correspond.

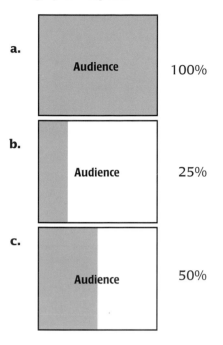

a. Audience — 100%

b. Audience — 25%

c. Audience — 50%

Materials Student Activity Sheet 1 (one per student)

Overview The context of this final story and the connected activities give students the opportunity to make their knowledge visible in drawings. Students show what percent of the seating area of a theater they think would be filled for various events by shading in a section of the diagram for each event.

About the Mathematics Percents that are relatively easy to estimate are known as *benchmark percents.* This final story contains most of the relevant things students should know about the key benchmark percents. Students should also get a feeling for the magnitude of 25 percent, 50 percent, and 100 percent.

Planning Have students work individually on their drawings for problem **11.** Then discuss their drawings in class. Stress the relationship between fractions and percents in discussing problem **11.**

Comments about the Problems

11. Informal Assessment This problem assesses students' ability to construct and use a visual model of a percent.

Several answers are possible. For example, if one student likes historical plays, he or she may expect a large audience. Also, students may choose fractions that cannot be expressed using easy benchmark percents. Encourage students to estimate the percent equivalents for such fractions.

12. The shaded diagrams below show the attendance at three other events. Estimate what part of the theater was filled for each event. Express each estimate as a fraction and as a percent.

a. A choral concert

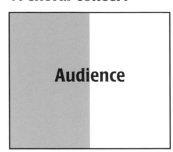

b. A jazz concert

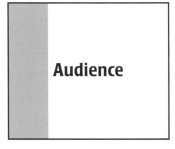

c. A rap concert

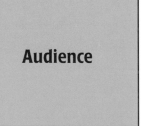

13. The middle school theater has about 300 seats. It is rather small compared to the downtown Civic Center theater, which can seat four times as many people. Last night both theaters had a 25 percent attendance rate. About how many people attended the events at each theater?

12. a. approximately $\frac{1}{2}$ or 50 percent

   b. approximately $\frac{1}{4}$ or 25 percent

   c. 1 or 100 percent

13. Since 25 percent of 300 is 75, 75 people attended the event at the middle school theater. Since 25 percent of 1,200 is 300, 300 people attended the event at the Civic Center theater. Accept percent estimates close to these answers.

Overview Students estimate what parts of the seating area of a theater are represented by the shaded areas in three diagrams. They express each answer as a fraction and as a percent.

About the Mathematics The focus of problem **13** is the relative nature of percent. The attendance at one auditorium differed from that of the other, while the percent of attendance (25 percent) is the same. The same percent does not necessarily mean the same absolute number, since a percent is calculated relative to a certain amount, or number.

Planning Students can work individually or in pairs on problems **12** and **13**.

Comments about the Problems

12. Informal Assessment This problem assesses students' ability to recognize the relationship between a fraction, a ratio, and a percent; their ability to estimate percents; and their ability to construct and use a visual model of percent.

   Discuss the relationship between the fraction and the percent in this problem.

13. This problem focuses on the relative nature of percents. Encourage students to use different strategies to solve the problem. It might be easier for some students to draw pictures and shade one-quarter of each to estimate the attendance at each theater.

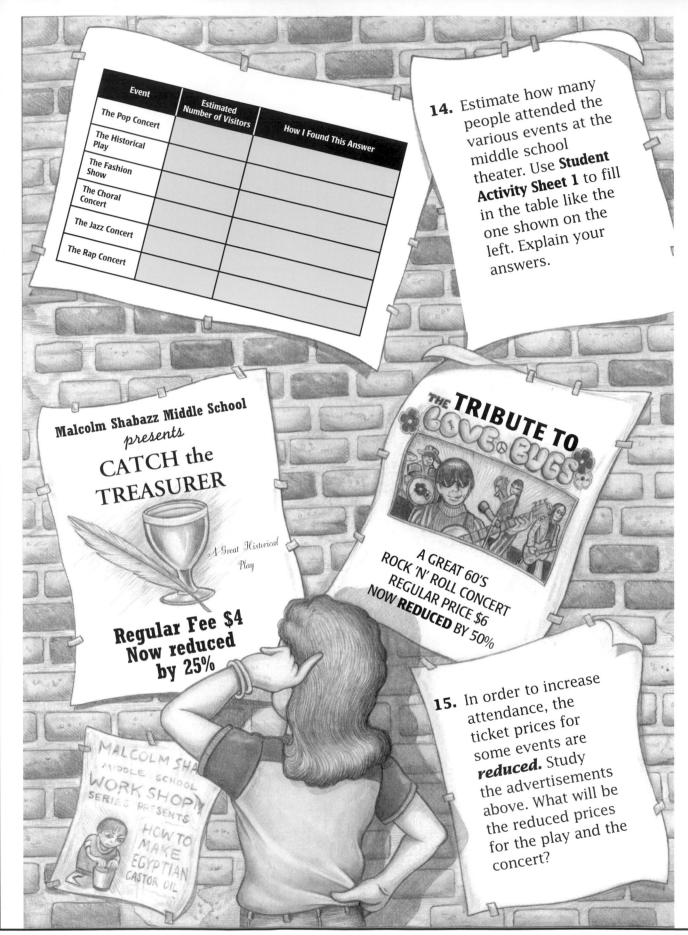

Event	Estimated Number of Visitors	How I Found This Answer
The Pop Concert		
The Historical Play		
The Fashion Show		
The Choral Concert		
The Jazz Concert		
The Rap Concert		

14. Estimate how many people attended the various events at the middle school theater. Use **Student Activity Sheet 1** to fill in the table like the one shown on the left. Explain your answers.

Malcolm Shabazz Middle School
presents
CATCH the
TREASURER

A Great Historical Play

**Regular Fee $4
Now reduced
by 25%**

MALCOLM SHA
MIDDLE SCHOOL
WORK SHOP
SERIES PRESENTS
HOW TO
MAKE
EGYPTIAN
CASTOR OIL

THE **TRIBUTE TO**
LOVE BUGS

A GREAT 60'S
ROCK 'N' ROLL CONCERT
REGULAR PRICE $6
NOW **REDUCED** BY 50%

15. In order to increase attendance, the ticket prices for some events are **reduced.** Study the advertisements above. What will be the reduced prices for the play and the concert?

14. Answers will vary. Here is a sample solution.

Event	Estimated Number of Visitors	How I Found This Answer
The Pop Concert	300 people	In problem 11, I shaded in the entire seating area of the theater. 100% of 300 = 300
The Historical Play	75 people	In problem 11, I shaded in one-fourth of the theater's seating area $\frac{1}{4}$ of 300 = 75
The Fashion Show	150 people	In problem 11, I shaded in one-half of the theater's seating area $\frac{1}{2}$ of 300 = 150
The Choral Concert	150 people	In problem 12, I estimated that one-half of the theater's seating area was filled $\frac{1}{2}$ of 300 = 150
The Jazz Concert	75 people	In problem 12, I estimated that one-fourth of the theater's seating area was filled $\frac{1}{4}$ of 300 = 75
The Rap Concert	300 people	In problem 12, I estimated that 100% of the theater's seating area was filled. 100% of 300 = 300

15. Price of the play, *Catch the Treasurer:* $3

Price of the concert, The Love Bugs: $3

Materials Student Activity Sheet 1 (one per student)

Overview Students estimate the number of people attending each of six events using their fraction and percent estimates from problems **11** and **12**. Students also find the reduced admission prices for two events.

About the Mathematics In these problems, simple percents are used as *operators* to find a percent estimate or an exact percent. It is not necessary to do real calculations since the benchmark fractions and percents in these problems are relatively easy to estimate.

Planning Students can work individually or in pairs on problems **14** and **15.**

Comments about the Problems

14. For the first three events, estimations should be based on the students' answers to problem **11.** Students who did not choose benchmark percents may need help in making their attendance estimates.

15. If students are having difficulty finding the reduced price for the rock concert, ask them: *What does 50 percent mean?* [50 out of 100, or 1 out of 2] *What fraction does it represent?* [$\frac{1}{2}$] This may help them understand percents as representations of fractions. Notice that the price is the same for both tickets, despite the different percent reductions.

Writing Opportunity Have students write about a favorite concert, play, or other event they have attended. Ask students to include an estimate of the attendance at the event and a drawing showing the percent of the seating area that was filled.

Summary

In this section, you read 10 stories that involve percents. You saw that fractions and percents can be used to represent the same thing. These stories were meant to show you what you might already know about percents.

Summary Questions

Use **Student Activity Sheet 2** to answer problems **16** and **17.**

16. Make drawings to express the following percents.

    a. 50 percent of the students are girls.

    b. 25 percent of the flowers are red.

    c. 100 percent of the cookies are broken.

17. Complete the drawing of the following situation.

On Monday, Sandy and June each bought the same kind of scarf. By Friday, strange things had happened. Sandy's scarf was 50 percent of its original length because it had shrunk in the dryer. June's scarf was 200 percent of its original length because her dog had stretched it.

16. Accept all reasonable drawings. Here are some examples of students' work.

17. Accept all reasonable drawings. Here is a sample student solution.

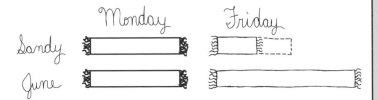

Materials Student Activity Sheet 2 (one per student)

Overview Students read and discuss the Summary, which reviews the main concepts covered in this section. Then students make drawings to represent most of the benchmark percents.

About the Mathematics Here are the five main concepts students should understand from this section if they are to succeed in the rest of the unit:

1. One-half ($\frac{1}{2}$) means the same thing as 50 percent of a given quantity.

2. One-quarter ($\frac{1}{4}$) means the same thing as 25 percent of a given quantity.

3. The total amount of a quantity means the same thing as 100 percent.

4. Doubling an amount means the same thing as 200 percent of the original amount.

5. Halving or dividing by two is equivalent to reducing a quantity by 50 percent.

Planning After students read the Summary, they may work individually or in pairs on problems **16** and **17.** It is important to discuss students' drawings so that the key points mentioned above are highlighted. After students complete Section A, you may assign appropriate activities from the Try This! section, located on pages 34–37 of the Student Book, as homework.

Comments about the Problems

16–17. Informal Assessment These problems can be used for homework and/or to assess the students' ability to use benchmark percents, to construct and use a visual model of percent, and to recognize the relationship between a fraction, a ratio and a percent.

The drawings may take any form (for instance, circles, rectangles, and so on). When reviewing these problems, discuss the students' drawings to emphasize that there are different ways to express percents.

Work Students Do

Students compare the degree to which different parking lots are occupied. They count the numbers of occupied and empty spaces and fill in a table for each lot. After completing the tables, students first shade in fraction bars to represent the portion of occupied spaces in each lot. Later, they use the data to shade in percent bars. Students then express the fraction of occupied spaces as a percent and compare the percents of occupied spaces for different parking lots. They also use ratio tables to find parking lots of different sizes that have the same fraction of occupied spaces.

Goals

Students will:

- estimate percents;
- use benchmark percents of 1%, 10%, 25%, and 50%;
- construct and use a visual model of percent;
- construct and use a ratio table to find what percent is equivalent to a given fraction or ratio, or vice versa;
- recognize the relationship between a fraction, a ratio, and a percent;
- understand the relative nature of percent;
- understand that percents are a means of standardizing to make comparisons, in static situations as well as in simple growth situations.

Pacing

- approximately four 45-minute sessions

Vocabulary

- ratio table
- percent bar

About the Mathematics

This section deals with comparing quantities with and without standardization. Percents are used to standardize quantitative relationships. For example, suppose you want to compare the survey results from two different-sized groups. The survey simply asks for a "yes" or "no" vote from participants. The number of "yes" votes must be expressed as a ratio of the total number of votes in each sample. The ratios can then be expressed as percents, which can be directly compared. Note that the word *relative* is not used in the Student Book and should not be mentioned yet. This concept is made explicit in Section C.

Visual models, such as the fraction/percent bars, are introduced here. The fraction/percent bars are used in the context of parking lots to estimate the fraction or percent of occupied spaces. Such models, used as estimation tools, will help students better understand the relative nature of percent.

Ratio tables are also informally introduced as a tool to compare situations with different possible totals. Both models are developed in the grade 5/6 unit *Some of the Parts.*

Materials

- Student Activity Sheets 3–8, pages 99–104 of the Teacher Guide (one of each per student)
- one-meter stick to use as model for the fraction/percent bar, page 31 of the Teacher Guide, optional (one per classroom)

Planning Instruction

You may want to introduce this section with a short discussion about parking lots. The questions on page 9 of the Student Book may help generate such a discussion. It is also advisable to have a class discussion about the multilevel parking garage of problem 19 to ensure students understand the total capacity of the parking garage during the different time periods.

Have students work together as a whole class on problems 1 and 19. Students may work in pairs or in small groups on problems 2–18 to discuss their ideas and to help each other find strategies to solve the problems. Students can work individually or in pairs on problems 20–21, and individually on problem 22.

There are no optional problems in this section. Discuss problems 2, 3, 4, 7, 10, 11, 14, and 15 with the whole class to ensure that all students understand the relationship between fractions and percents.

Homework

You can assign problems 11–14 (page 30 of the Teacher Guide), problem 19 (page 34 of the Teacher Guide), problems 20–21 (page 36 of the Teacher Guide), the Extensions (pages 29 and 39 of the Teacher Guide), and the Writing Opportunity (page 39 of the Teacher Guide). After students complete Section B, you may assign appropriate activities from the Try This! section located on pages 34–37 of the *Per Sense* Student Book. These activities can be used to reinforce the main concepts of Section B.

Planning Assessment

- Problem 14 can be used to informally assess students' ability to construct and use a ratio table to find what percent is equivalent to a given fraction or ratio, or vice versa.
- Problem 20 can be used to informally assess students' ability to construct and use a visual model of percent; to recognize the relationship between a fraction, a ratio, and a percent; to estimate percents; and to understand that percents are a means of standardizing to make comparisons, in static situations as well as in simple growth situations.
- Problem 21 can be used to informally assess students' understanding of the relative nature of percent. It also shows their ability to use benchmark percents and to recognize the relationship between a fraction, a ratio, and a percent.
- Problem 22 can be used to informally assess students' ability to construct and use a visual model of a percent; to recognize the relationship between a fraction, a ratio, and a percent; and to estimate percents.

B. USING PERCENTS TO COMPARE

Lots of Luck

1. The parking lot in the photo is very popular. How can you tell?

2. Do you think the parking lot below is full? Explain your answer.

3. Compare the parking lot in the photo with the one below. Which is better and why?

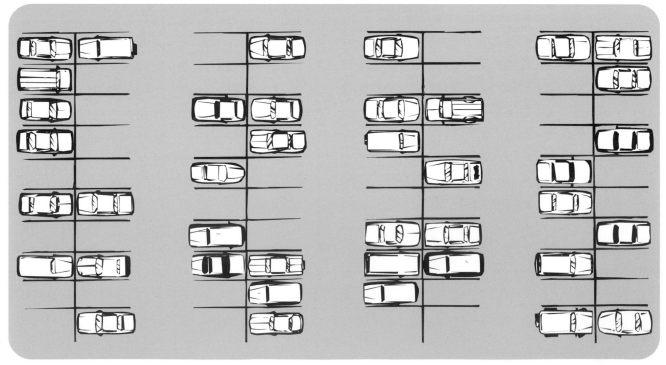

Solutions and Samples
of student work

1. Answers will vary. Accept any well-defended answer. This is meant only as an introduction to this section.

2. Answers will vary. Students who compare the number of occupied spaces to the total spaces might respond that the parking lot is half full, or 50 percent full, because 40 spaces out of 80 are occupied. Other students might visually estimate that the lot is about half full.

3. Answers will vary. The parking lot in the photo might be considered "better" if it is located closer to your destination or if it is less expensive than the one at the bottom of the page. The parking lot at the bottom of the page might be considered "better" if it is closer to your destination, if it and the other lot are equidistant from your destination, if it is less expensive or the same price as the other lot, or because it allows you to drive in and out more easily.

Hints and Comments

Overview Students read and discuss three problems that introduce the context of this section: parking lots.

About the Mathematics In this section, fractions and ratios are used to describe and compare since they tap into students' mathematical concepts of the relationships among quantities. Percents are only a standardization of these concepts in terms of parts per hundred.

The context of the parking lots is used to help students get a feeling for the *relative nature* of percents. It is used here as an estimation tool. The *benchmark percents* and *fractions* from Section A play an important role in these estimations. Within the context of the parking lots, the fraction/percent model is informally introduced.

Planning You may use problem **1** for a short introduction to the context of parking lots. Students may then work on problems **2** and **3** in pairs or in small groups. Then, discuss these problems with the whole class.

Comments about the Problems

1. At first glance, students may notice that this parking lot leaves no access for cars in the middle to exit the lot. They are hemmed in on all sides. Ask students: *What kind of parking lot might this be?* [It could be a police impoundment yard or a lot near a factory where new cars are stored.] Finally, you can informally ask about the relative fullness of the parking lot.

2. The answers may have different levels of accuracy. Some students may compare the number of occupied spaces to the number of empty spaces. Other students may compare the number of empty spaces to the total capacity. [In either case, the lot is about 50 percent full or 50 percent empty.] Other students may explain that the lot is not full because there are still some empty spaces.

3. This problem should raise issues such as what the different purposes of the lots might be, which lot is more accessible, and so on. For car storage, the lot in the photo might be "better" because it has a bigger capacity. For normal parking purposes, the lot below might be "better" for drivers looking for easy access in and out of the lot.

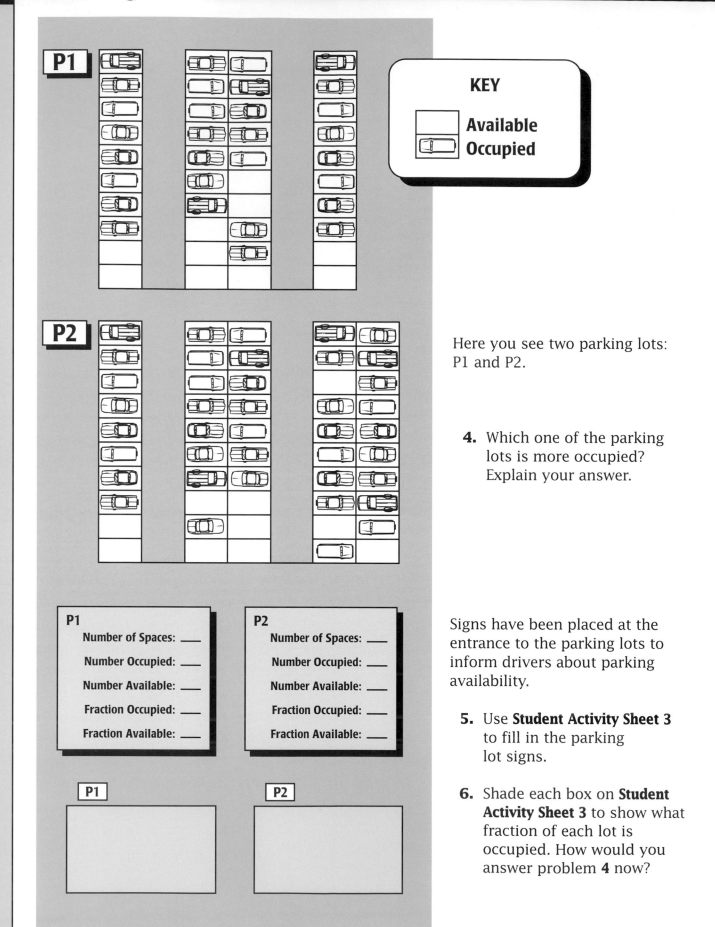

KEY

☐ Available

☐ Occupied

Here you see two parking lots: P1 and P2.

4. Which one of the parking lots is more occupied? Explain your answer.

P1

Number of Spaces: ___

Number Occupied: ___

Number Available: ___

Fraction Occupied: ___

Fraction Available: ___

P2

Number of Spaces: ___

Number Occupied: ___

Number Available: ___

Fraction Occupied: ___

Fraction Available: ___

P1

P2

Signs have been placed at the entrance to the parking lots to inform drivers about parking availability.

5. Use **Student Activity Sheet 3** to fill in the parking lot signs.

6. Shade each box on **Student Activity Sheet 3** to show what fraction of each lot is occupied. How would you answer problem **4** now?

4. Lot P2 is more occupied than Lot P1. Lot P2 has 40 out of 50 spaces occupied (80 percent occupied), and Lot P1 has only 30 out of 40 spaces occupied (75 percent occupied). Students who compare only the numbers of empty spaces in both lots may incorrectly conclude that both lots are equally occupied since they each have 10 empty spaces.

5.

P1	
Number of Spaces:	40
Number Occupied:	30
Number Available:	10
Fraction Occupied:	$\frac{30}{40}$ or $\frac{3}{4}$
Fraction Available:	$\frac{10}{40}$ or $\frac{1}{4}$

P2	
Number of Spaces:	50
Number Occupied:	40
Number Available:	10
Fraction Occupied:	$\frac{40}{50}$ or $\frac{4}{5}$
Fraction Available:	$\frac{10}{50}$ or $\frac{1}{5}$

6. The box for Lot P1 should be shaded to show that three-fourths of the lot is occupied. The box for Lot P2 should be shaded to show that four-fifths of the lot is occupied. Here are some examples showing different levels of understanding:

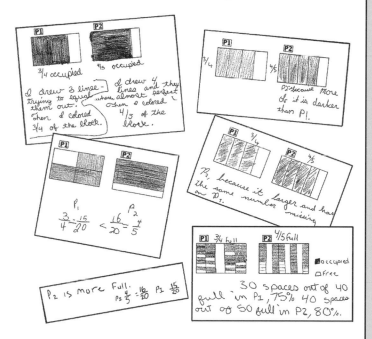

Materials Student Activity Sheet 3 (one per student)

Overview Students compare the fractions of occupied spaces of two parking lots. They count the numbers of occupied and unoccupied spaces and use these numbers to fill in the parking signs. They also express the parts of the lot that are occupied and unoccupied as fractions. Students then shade in fraction models to represent the fraction of occupied spaces in each lot.

About the Mathematics Quantities can be compared absolutely or relatively. The relevancy of each type of comparison depends on the specific situation. Although either comparison method can be used in this context, only a relative comparison will give information regarding the percents of occupancy of the parking lots. The fraction bar model is preceded by a shaded box in problem **6.** At this point, the visual impact of the fraction is important. Later, numbers and percents will be added to the fraction models to show the relationship between fractions and percents. Being able to draw and interpret fraction models will be key to students' success in the rest of the unit.

Planning Students may work in pairs or small groups on problems **4–6.** After they have finished problem **4,** discuss it with the whole class. After a brief discussion of their responses and logic, have them proceed with problem **5.**

Comments about the Problems

4. Since each parking lot has 10 empty spaces, students must find ways to express the number of occupied spaces relative to the total number of spaces. Some students may use ratios or fractions, while others may show their reasoning in an explanation. Do not stress using fractions here; fractions will be the focus of the following problems.

5. If students are having difficulty, you may want to return to this problem after students have finished problem **6.** Some students may need to review the fraction concepts in the grade 5/6 unit *Some of the Parts.*

6. Accept all student drawings as long as the shaded fraction models correctly represent the fraction of occupied spaces in each lot.

Here are two more parking lots: P3 and P4.

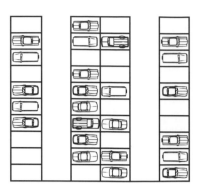

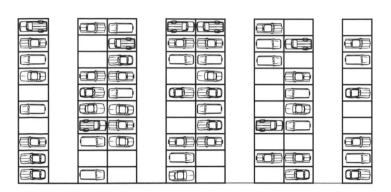

7. At first glance, which lot do you think is more occupied? Explain.

Use **Student Activity Sheet 4** to solve the following problems.

8. Complete the parking lot signs for P3 and P4.

You can also make bars to show the numbers of cars in the lots.

9. Shade each bar to show the number of cars in each lot.

P3

Number of Spaces: ___

Number Occupied: ___

Number Available: ___

Fraction Occupied: ___

Fraction Available: ___

P4

Number of Spaces: ___

Number Occupied: ___

Number Available: ___

Fraction Occupied: ___

Fraction Available: ___

P3 — 0 cars .... 40 cars

P4 — 0 cars .... 80 cars

10. Fill in the blanks in the first table so that parking lots A, B, and C have the same fraction of occupied spaces as P3.

Then fill in the second table so that parking lots D, E, and F have the same fraction of occupied spaces as P4.

This kind of table is called a ***ratio table.*** Why do you think it is called this?

	P3	Lot A	Lot B	Lot C
Spaces Occupied	24		36	
Total Spaces	40	20		10

	P4	Lot D	Lot E	Lot F
Spaces Occupied	56	28		35
Total Spaces	80		10	

7. Answers will vary. Because the percents of occupancy for both lots are close, accept either answer. Lot P4 is more occupied (56 out of 80, $\frac{7}{10}$, or 70 percent of the spaces are occupied) than Lot P3 (24 out of 40, $\frac{3}{5}$, or 60 percent of the spaces are occupied).

8.

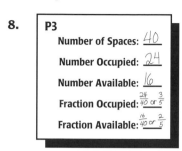

P3
Number of Spaces: 40
Number Occupied: 24
Number Available: 16
Fraction Occupied: $\frac{24}{40}$ or $\frac{3}{5}$
Fraction Available: $\frac{16}{40}$ or $\frac{2}{5}$

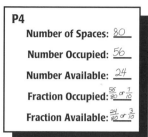

P4
Number of Spaces: 80
Number Occupied: 56
Number Available: 24
Fraction Occupied: $\frac{56}{80}$ or $\frac{7}{10}$
Fraction Available: $\frac{24}{80}$ or $\frac{3}{10}$

9.

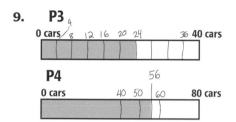

P3

0 cars 4 8 12 16 20 24 36 40 cars

P4

0 cars 40 50 56 60 80 cars

10. See ratio tables below. Possible explanation: This is called a ratio table because, for all columns, the numbers in the top row have the same ratio to the numbers in the bottom row.

	P3	Lot A	Lot B	Lot C
Spaces Occupied	24	12	36	6
Total Spaces	40	20	60	10

$\div 2$ $\times 3$ $\div 6$

The number of total spaces for Lot B can be found either by multiplying the number of total spaces for Lot A by three or by adding the number of total spaces for Lots P3 and A.

	P4	Lot D	Lot E	Lot F
Spaces Occupied	56	28	7	35
Total Spaces	80	40	10	50

$\div 2$ $\div 4$ $\times 5$

The number of total spaces for Lot F can be found either by multiplying the number of total spaces for Lot E by five or by adding the number of total spaces for Lots D and E.

Materials Student Activity Sheet 4 (one per student)

Overview Students compare the fractions of occupied spaces in two more parking lots and shade in bars to show the number of cars in each lot. Students also use a *ratio table* to create parking lots of different sizes that have the same fraction of occupied to total spaces.

About the Mathematics The shaded boxes on the previous page now take the shape of a bar, introducing the *fraction bar* model. The fraction bar is a visual model that builds on students' ideas of part-whole relationships.

Another visual model, the *ratio table,* first appears here. With a ratio table, equivalent ratios can be found by means of halving, doubling, multiplying, dividing, adding, or subtracting. Both models are developed in the grade 5/6 unit *Some of the Parts.*

Planning Students may work on problems **7–10** in pairs or in small groups. Discuss problems **7** and **10** with the whole class. The Extension activity below may be assigned as homework.

Comments about the Problems

7. Students should be able to discuss how they arrived at their answers. If students are having difficulty, have them refer back to problem **4.**

8. Observe the class to see whether or not all students now understand how to complete the data in the parking lot signs.

9. An exact answer is not necessary at this point. However, students should be able to give a reasonable estimate of the shaded part in each fraction bar. Several strategies can be used: repeated halving, finding a familiar fraction and then writing multiples of that number, and closely examining the visual model.

10. Do not stress mechanical methods for finding equivalent fractions. Students need to struggle a bit to get a feel for what these fractions represent.

Extension You may ask students to create parking lots of their own. Have them sketch the layout of their parking lot and draw cars to indicate each occupied space. Then tell students to make and fill in a parking lot sign as they did in problems **5** and **8.**

Here are parking lots P5 and P6.

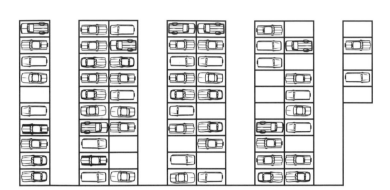

P5   **P6**

11. Which parking lot is more occupied? Explain.

P5

Number of Spaces: ___

Number Occupied: ___

Number Available: ___

Fraction Occupied: ___

Fraction Available: ___

P6

Number of Spaces: ___

Number Occupied: ___

Number Available: ___

Fraction Occupied: ___

Fraction Available: ___

0 cars   40 cars

P5

0 cars   75 cars

P6

	P5	Lot G	Lot H	Lot I
Spaces Occupied	36		9	
Total Spaces	40	20		60

	P6	Lot J	Lot K	Lot L
Spaces Occupied	60			
Total Spaces	75	25	50	100

Use **Student Activity Sheet 5** to solve problems **12–14.**

12. Fill in the parking lot signs.

13. Shade each bar to show what fraction of each lot is occupied.

14. Fill in the blanks in the first table so that parking lots G, H, and I have the same fraction of occupied spaces as P5.

Then fill in the second table so that parking lots J, K, and L have the same fraction of occupied spaces as P6.

11. Lot P5 is more occupied. Possible explanations: Lot P5 has 36 out of 40 occupied spaces while Lot P6 has 60 out of 75 occupied spaces ($\frac{36}{40}$, or $\frac{9}{10}$, is greater than $\frac{60}{75}$, or $\frac{4}{5}$). Lot P5 has only 4 empty spaces out of 40 while Lot P6 has 15 empty spaces out of 75 ($\frac{4}{40}$, or $\frac{1}{10}$, is less than $\frac{15}{75}$, or $\frac{1}{5}$).

12.

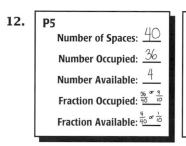

P5		P6	
Number of Spaces:	40	Number of Spaces:	75
Number Occupied:	36	Number Occupied:	60
Number Available:	4	Number Available:	15
Fraction Occupied:	$\frac{36}{40}$ or $\frac{9}{10}$	Fraction Occupied:	$\frac{60}{75}$ or $\frac{4}{5}$
Fraction Available:	$\frac{4}{40}$ or $\frac{1}{10}$	Fraction Available:	$\frac{15}{75}$ or $\frac{1}{5}$

13.

P5
0 cars   36   40 cars

P6
0 cars   60   75 cars

14.

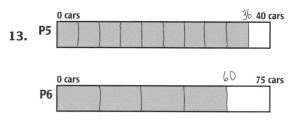

	P5	Lot G	Lot H	Lot I
Spaces Occupied	36	18	9	54
Total Spaces	40	20	10	60

÷ 2   ÷ 2   × 6

The number of total spaces for Lot H can be found either by dividing the number of total spaces for Lot G by two or by dividing the number of total spaces for P5 by four.

The number of occupied spaces for Lot I can be found either by multiplying the number of occupied spaces for Lot G by three or by adding the numbers of occupied spaces for Lots P5 and G.

÷ 3   × 2   × 2

	P6	Lot J	Lot K	Lot L
Spaces Occupied	60	20	40	80
Total Spaces	75	25	50	100

The number of occupied spaces for Lot K can be found either by doubling the number of occupied spaces of Lot J or by subtracting the number of occupied spaces of Lot J from that of Lot P6.

The number of occupied spaces for Lot L can be found by doubling the number of occupied spaces of Lot K or by adding the number of occupied spaces of Lots P6 and J.

Materials Student Activity Sheet 5 (one per student); one-meter stick, optional (one per classroom)

Overview Students compare the fractions of occupied spaces in two more parking lots. These problems are similar to those on the previous page. You may want to use a one-meter stick to demonstrate the fraction/percent bar model with the whole class.

About the Mathematics The fraction/percent bar model supports estimations and calculations with percent. For example, suppose Susan brought four-dozen cookies to class, and the class ate three-fourths of the cookies. How many cookies were eaten?

0	12	24	36	48
0%	25%	50%	75%	100%

The numbers on the top of the bar represent the number of cookies. The numbers on the bottom of the bar represent the percent of the cookies eaten. By shading the fraction of cookies that were eaten, the students can read the number of cookies eaten (36), the fraction of cookies eaten ($\frac{3}{4}$), and the percent of cookies eaten (75 percent).

Planning Students may work in pairs or in small groups on problems **11–14**. If students did not have difficulty with the problems on the previous page, you can assign these problems as homework. Be sure to discuss problems **11** and **14** with the whole class.

Comments about the Problems

11. Homework This problem may be assigned as homework. Have students share the strategies they used to obtain their answers.

12. Homework This problem may be assigned as homework. At this point, students should be able to complete the data for the parking lot signs.

13. Students should now be able to provide a reasonable estimate of the shaded part. Several strategies can be used: repeated halving, finding a familiar fraction and writing multiples of that number, and so forth.

14. Informal Assessment This problem assesses students' ability to construct and use a ratio table to find what percent is equivalent to a given fraction or ratio, or vice versa. This problem may also be assigned as homework.

Discuss the different strategies that students used to find the missing numbers in the ratio tables. Note that adding or subtracting in the ratio tables can also be a useful strategy.

Decisions, Decisions...

Parking lots P1–P6 belong to one company. The company managers want to know which parking lots are often completely or almost completely occupied. The managers are thinking about expanding these lots. The lots that are often nearly empty might be closed.

15. a. Using the parking lot signs for lots P1–P6 from **Student Activity Sheets 3, 4,** and **5,** how might the managers determine which lots should be expanded or closed?

b. Which parking lot(s) would you choose to close?

c. Which one(s) would you choose to expand?

P1

0   40
0%   100%

P2

0   50
0%   100%

One of the managers compared the records of the six parking lots from the same day and time. The comparisons are shown on the left.

16. Use the shaded bars to find what fraction of each lot is occupied.

P3

0   40
0%   100%

17. Express what part of each lot is occupied using a percent. Use the bars to find your answers.

P4

0   80
0%   100%

18. a. Based on this information, which parking lot(s) would you choose to close or expand?

P5

0   40
0%   100%

b. Compare your answers to those you got in problems **15b** and **c.**

P6

0   75
0%   100%

c. Is this a good way to decide which lots to close or expand?

15. a. Answers will vary. The managers may choose certain lots to be expanded because they have few empty spaces or because certain lots are larger and fairly full. The managers may choose to close certain lots because they have more empty spaces or because they are smaller and do not have as many cars.

b. Answers will vary. Some students may choose to close lots P3, P4, and P6 because they have more empty spaces. Accept any answer justified with logical reasoning.

c. Answers will vary. Some students may choose to expand lots P1, P2, and P5 because they have few empty spaces. Other students may choose to expand lots P4 and P6 because they are larger and fairly full. Accept any answer justified with logical reasoning.

16. $P1 = \frac{30}{40}$ or $\frac{3}{4}$; $P2 = \frac{40}{50}$ or $\frac{4}{5}$; $P3 = \frac{24}{40}$ or $\frac{3}{5}$; $P4 = \frac{56}{80}$ or $\frac{7}{10}$; $P5 = \frac{36}{40}$ or $\frac{9}{10}$; $P6 = \frac{60}{75}$ or $\frac{4}{5}$.

17.

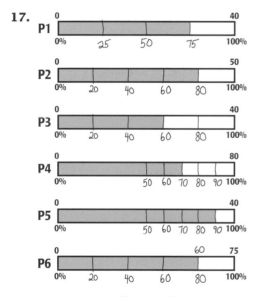

18. a. Answers will vary. The ideal lots to expand are P2, P5, and P6. P3 might be closed.

b. Answers will depend on student responses for problems **15b** and **c.** Some students may have greatly revised the original list of suggested lots to expand or to close.

c. Answers will vary. A student might say, "This is not a good way to decide which lots to expand or close because the data are from only one day and time."

Materials Student Activity Sheets 3–5 (one of each per student)

Overview Students compare the data and percent bars for lots P1–P6 to decide which lots might be expanded and which lots might be closed.

About the Mathematics The fraction/percent bar model is useful in making relative comparisons between different quantities. Since the bars are already shaded, this comparison can be done immediately. The absolute numbers also play a role in making the decisions regarding which lots to expand or close.

After students have expressed the occupied portion of each lot using a fraction and a percent, the number of occupied spaces can be found by using the fractions as operators.

Planning Students may work on problems **15–18** in small groups. After students finish problem **15,** discuss their answers and reasoning with the whole class.

Comments about the Problems

15. See if students spontaneously use percents to compare the six lots. If not, you do not need to explain the use of percents now. It will be dealt with in the next problem. The decision to expand or close each lot should not be based solely on quantitative information. Additional factors such as available space, usage, and the number of spaces needed will determine whether a lot is expanded or closed. Be sure to discuss students' answers and reasoning with the whole class.

16–17. As students work with the fraction/percent bars now, observe how well they adapt to the percent line on the bottom of the bar. If they are having difficulty, relate the bars to benchmark fractions such as $\frac{1}{2}, \frac{1}{3}, \frac{1}{4}, \frac{1}{5}$, and so forth.

18. Students may or may not change their answers to problem **15.** Be sure that students are making their decisions based on the percent bars.

WORKING ON MANY
Levels

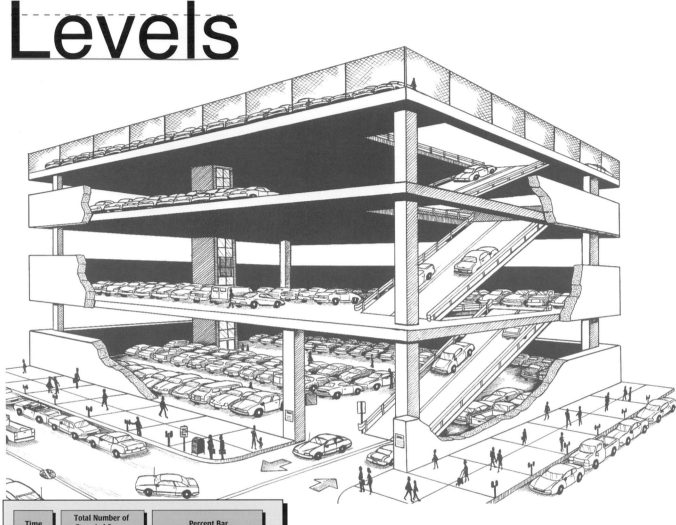

Time	Total Number of Occupied Spaces	Percent Bar
8 A.M.–10 A.M.	90	0% no cars 100% full
10 A.M.–12 P.M.	120	
12 P.M.–2 P.M.	180	
2 P.M.–4 P.M.	240	
4 P.M.–6 P.M.	300	
6 P.M.–8 P.M.	320	
8 P.M.–10 P.M.	400	
10 P.M.–12 A.M.	400	
12 A.M.–2 A.M.	100	
2 A.M.–4 A.M.	50	
4 A.M.–6 A.M.	40	
6 A.M.–8 A.M.	60	

This multilevel parking lot is open 24 hours a day. It has four levels, and each level has 100 parking spaces.

During office hours (from 8:00 A.M. to 6:00 P.M.), the lower level is reserved for employees of a bank. As a result, there are only 300 spaces available for the general public during these hours.

19. Shade the parking lot bars on **Student Activity Sheet 6** to show what percent of the parking lot is occupied during each two-hour period.

19. See shaded bars below.

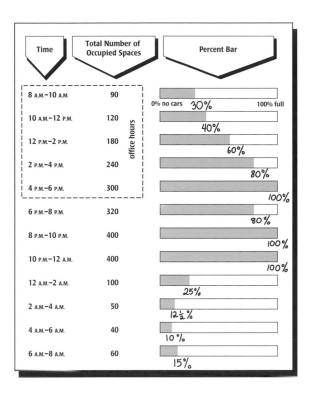

Materials Student Activity Sheet 6 (one per student)

Overview Students read about a multilevel parking lot. They use information about the number of occupied spaces and the total number of available parking spaces to shade in a percent bar for each time period.

About the Mathematics Percent bars are used again to express the part-whole relationship between the number of occupied spaces and the total number of spaces. The numbers used in this problem can be expressed as simple fractions in most cases. However, since so many time intervals are involved, it is easier to use percents instead of fractions to make a relative comparison. When the percent bars are completed, have students turn their activity sheet to a horizontal position. The percent bars then become bar graphs similar to the graphs students encountered in the grade 5/6 unit *Picturing Numbers.*

Planning You may want to have a short class discussion to introduce the multilevel parking garage context here. You might ask students what percents they recognize at first glance. For example, *For what time period(s) is the lot 100 percent full?* [Between 4 P.M. and 6 P.M. and between 8 P.M. and 12 A.M.] After this short introduction, you may have students finish the problem at home, since it takes a long time to complete.

Comments about the Problems

19. Homework This problem can be assigned as homework. Remind students that between 8 A.M. and 6 P.M., there are only 300 total parking spaces. After 6 P.M., there are 400 total parking spaces.

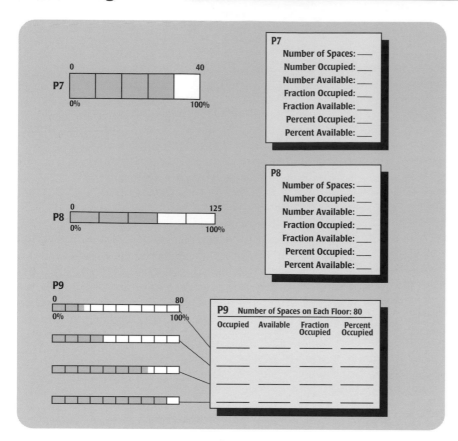

P7

Number of Spaces: ——	
Number Occupied: ___	
Number Available: ___	
Fraction Occupied: ___	
Fraction Available: ___	
Percent Occupied: ___	
Percent Available: ___	

P8

Number of Spaces: ——	
Number Occupied: ___	
Number Available: ___	
Fraction Occupied: ___	
Fraction Available: ___	
Percent Occupied: ___	
Percent Available: ___	

P9 Number of Spaces on Each Floor: 80

Occupied	Available	Fraction Occupied	Percent Occupied
___	___	___	___
___	___	___	___
___	___	___	___
___	___	___	___

Use **Student Activity Sheet 7** to solve the following problem.

20. Here are more bars, for parking lots P7, P8, and P9. What fraction of each lot or floor is occupied? Express each fraction as a percent. Then complete the parking lot signs.

P11

Number of Spaces: ____	Number of Spaces: ____	Number of Spaces: ____
Number Occupied: ____	Number Occupied: ____	Number Occupied: ____
Number Available: ____	Number Available: ____	Number Available: ____
Percent Occupied: _50_	Percent Occupied: _50_	Percent Occupied: _50_
Percent Available: ____	Percent Available: ____	Percent Available: ____

P12

Number of Spaces: ____	Number of Spaces: ____	Number of Spaces: ____
Number Occupied: ____	Number Occupied: ____	Number Occupied: ____
Number Available: ____	Number Available: ____	Number Available: ____
Percent Occupied: _25_	Percent Occupied: _25_	Percent Occupied: _25_
Percent Available: ____	Percent Available: ____	Percent Available: ____

P13

Number of Spaces: ____	Number of Spaces: ____	Number of Spaces: ____
Number Occupied: ____	Number Occupied: ____	Number Occupied: ____
Number Available: ____	Number Available: ____	Number Available: ____
Percent Occupied: _10_	Percent Occupied: _10_	Percent Occupied: _10_
Percent Available: ____	Percent Available: ____	Percent Available: ____

Use **Student Activity Sheet 8** to solve the following problem.

21. On the left are three signs for each parking lot. Fill them in with different numbers of occupied and available spaces so that for each parking lot, all three signs show the same *percent* of occupied spaces. You may use a parking lot bar to help you.

20. Lot P7 is four-fifths (80 percent) occupied. Lot P8 is three-fifths (60 percent) occupied. The first floor of Lot P9 is one-fourth (25 percent) occupied. The second floor of P9 is two-fifths (40 percent) occupied. The third floor of P9 is three-fourths (75 percent) occupied and the fourth floor of P9 is nine-tenths (90 percent) occupied. See sample percent bars below.

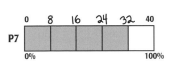

P7
Number of Spaces: $\underline{40}$
Number Occupied: $\underline{32}$
Number Available: $\underline{8}$
Fraction Occupied: $\underline{\frac{4}{5}}$
Fraction Available: $\underline{\frac{1}{5}}$
Percent Occupied: $\underline{80\%}$
Percent Available: $\underline{20\%}$

P8
Number of Spaces: $\underline{125}$
Number Occupied: $\underline{75}$
Number Available: $\underline{50}$
Fraction Occupied: $\underline{\frac{3}{5}}$
Fraction Available: $\underline{\frac{2}{5}}$
Percent Occupied: $\underline{60\%}$
Percent Available: $\underline{40\%}$

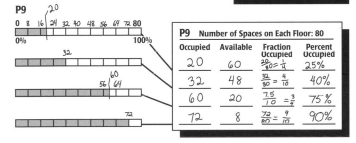

P9 Number of Spaces on Each Floor: 80

Occupied	Available	Fraction Occupied	Percent Occupied
20	60	$\frac{20}{80} = \frac{1}{4}$	25%
32	48	$\frac{32}{80} = \frac{4}{10}$	40%
60	20	$\frac{7.5}{10} = \frac{3}{4}$	75%
72	8	$\frac{72}{80} = \frac{9}{10}$	90%

21. Answers will vary. Possible answers:

P11
Number of Spaces: $\underline{100}$
Number Occupied: $\underline{50}$
Number Available: $\underline{50}$
Percent Occupied: $\underline{50}$
Percent Available: $\underline{50}$

P11
Number of Spaces: $\underline{200}$
Number Occupied: $\underline{100}$
Number Available: $\underline{100}$
Percent Occupied: $\underline{50}$
Percent Available: $\underline{50}$

P11
Number of Spaces: $\underline{50}$
Number Occupied: $\underline{25}$
Number Available: $\underline{25}$
Percent Occupied: $\underline{50}$
Percent Available: $\underline{50}$

P12
Number of Spaces: $\underline{100}$
Number Occupied: $\underline{25}$
Number Available: $\underline{75}$
Percent Occupied: $\underline{25}$
Percent Available: $\underline{75}$

P12
Number of Spaces: $\underline{200}$
Number Occupied: $\underline{50}$
Number Available: $\underline{150}$
Percent Occupied: $\underline{25}$
Percent Available: $\underline{75}$

P12
Number of Spaces: $\underline{80}$
Number Occupied: $\underline{20}$
Number Available: $\underline{60}$
Percent Occupied: $\underline{25}$
Percent Available: $\underline{75}$

P13
Number of Spaces: $\underline{100}$
Number Occupied: $\underline{10}$
Number Available: $\underline{90}$
Percent Occupied: $\underline{10}$
Percent Available: $\underline{90}$

P13
Number of Spaces: $\underline{200}$
Number Occupied: $\underline{20}$
Number Available: $\underline{180}$
Percent Occupied: $\underline{10}$
Percent Available: $\underline{90}$

P13
Number of Spaces: $\underline{50}$
Number Occupied: $\underline{5}$
Number Available: $\underline{45}$
Percent Occupied: $\underline{10}$
Percent Available: $\underline{90}$

Materials Student Activity Sheets 7 and 8 (one of each per student)

Overview Again, students use percent bars and express the shaded parts as fractions and percents. Then they use the percent bars to find information such as the numbers of total, occupied, and empty spaces for each parking lot. Finally, students supply numbers for parking lot signs so that the data in each sign represent the same percent of occupied spaces.

Planning Students may work individually or in pairs on problems **20–21.** These problems can be assigned as homework and/or used as informal assessment.

Comments about the Problems

20. Informal Assessment This problem can be used to informally assess students' ability to estimate percents; construct and use a visual model of percent; and recognize the relationship between a fraction, a ratio, and a percent. It also shows whether or not they understand that percents are a means of standardizing to make comparisons, in static situations as well as in simple growth situations.

Different strategies are possible. If students are having difficulty recognizing the fractions, have them label the bars with the numbers of total spaces and then reason about the fractions and, subsequently, the numbers of cars in the lots. Stress the relationship between the fractions and percents. Make sure that students use the pictures with the problems.

21. Informal Assessment This problem assesses students' understanding of the relative nature of percents. It also shows their ability to use benchmark percents and to recognize the relationship between a fraction, a ratio, and a percent.

If students are having difficulty, remind them of the relationship between the fractions and the percents and have them draw parking lot bars.

Summary

In this section, you saw how **_percent bars_** can be used to compare different situations.

0   80

0%   100%

The percent bar can be used to find parts of a whole, expressed in percents. A fully shaded bar is 100 percent. A bar without shading is 0 percent.

The bar above shows 25 percent of 80.

I FIGURED OUT EACH OF THESE PERCENTS DIFFERENTLY.

Tonya

Summary Question

Tonya had to solve this problem on a test. What percent is the first number of the second?

    a. 60 out of 80

    b. 50 out of 85

    c. 36 out of 40

Here are the drawings Tonya made for each solution.

a.
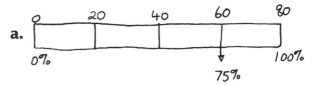

b.
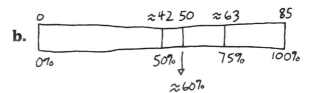

c.
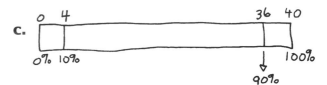

22. Explain how Tonya solved each problem.

22. There are three basic strategies that Tonya used.

a. Tonya may have recognized an easy fraction ($\frac{60}{80} = \frac{3}{4}$) and known the percent (75 percent) that the fraction represented.

b. Tonya may have made an estimation by first marking easy fractions, and then approximating the answer with smaller steps (50 out of 85 is closer to $\frac{1}{2}$ than $\frac{3}{4}$, so it is about 60 percent).

c. Tonya appears to have used another known percent to calculate the final percent. She may have found 10 percent of 40 (4) and then recognized that 36 is nine times that 10 percent, or 90 percent.

Overview Students read and discuss the Summary, which reviews the main concepts of this section. Then they explain the different strategies that Tonya used to express ratios as percents by interpreting the percent bar drawings she made.

Planning Students may work individually on problem **22.** This problem can be used as an informal assessment and/or homework. You can also use the Extension below as informal assessment. The Extension activity below may also be assigned as homework. After students complete Section B, you may assign appropriate activities from the Try This! section, located on pages 34–37 of the Student Book, as homework.

Comments about the Problems

22. Informal Assessment This problem can be used to informally assess students' ability to construct and use a visual model of a percent; recognize the relationship between a fraction, a ratio, and a percent; and estimate percents. The goal of this activity is to help students retain the strategies and concepts they have learned in this section. The fraction/percent bar strategies are very important. However, students should be encouraged to develop their own strategies.

Extension Here is an activity that can be used as an informal assessment or homework assignment. Write pairs of questions that ask students to express a ratio as a fraction and as a percent, and then to draw a picture to "prove" that their answers (or reasonable estimates) are correct. Here is an example:

a. Express 15 out of 30 as a fraction and a percent.

b. Draw a picture that "proves" your answers to problem **a.**

Vary the questions. You can give these short informal assessments or homework assignments periodically throughout the unit to assess students' understanding of the relationship between fractions and percents.

Writing Opportunity You may ask students to write their answers to problem **22** in their journals.

SECTION C. BENCHMARK PERCENTS

Work Students Do

In the context of a baseball game, students estimate the numbers of Giants and Dodgers fans at one game. They describe these numbers using fractions and percents. From given ratios of Giants fans and Dodgers fans, students find the numbers and percents of fans using different percent strategies. To rank the numbers of Tigers fans at different games, students first look at the actual numbers of Tigers fans and then at the numbers of Tigers fans in relation to the total number of spectators.

Students then read and interpret graphs showing the percents of fans in the stadium during different time intervals. Finally, they interpret the results of two polls, using different strategies to find percents.

Goals

Students will:

- estimate percents;
- use the benchmark percents of 1%, 10%, 25%, and 50%;
- construct and use a ratio table to find what percent is equivalent to a given fraction or ratio, or vice versa;
- construct and use a visual model of percent;
- recognize the relationship between a fraction, a ratio, and a percent;
- understand the relative nature of percent;
- understand that fractions, ratios, and percents are used as comparison tools;
- determine which strategy for finding the percent is most appropriate in a given situation;
- determine whether or not percents are used appropriately in a decision-making situation.

Pacing

- approximately four 45-minute class sessions

Vocabulary

- benchmark percent
- graph
- poll
- survey

About the Mathematics

The different strategies for calculating percent are dealt with in greater depth in this section. Topics include calculating percent by using a ratio table, the global calculation of percent, predicting large numbers using percent, taking samples, making comparisons, and reading graphs. In the summary activity, the strategy of finding easily recognizable percents (such as 1 percent, 10 percent, 25 percent, and 50 percent) and using these benchmark percents in combination to estimate a more difficult quantity is developed.

Materials

• Student Activity Sheets 9–12, pages 105–108 of the Teacher Guide (one of each per student)

Planning Instruction

You may want to use problem 1 to introduce the context of this section, baseball teams and stadiums, in a class discussion.

Students can work on problems 1–7 and 9–26 in pairs or in small groups to give them the opportunity to cooperate and discuss their strategies. Students may work individually on problems 8 and 27.

Problems 23–26 are optional. If time is a concern, these problems may be skipped. Be sure to discuss problems 4, 6, 7, 9–11, and 15, as well as the strategies in the Summary, with the whole class. Emphasize students' strategies in these discussions.

Homework

You may assign problems 23–26 (page 58 of the Teacher Guide) for homework. After students complete Section C, you may assign appropriate activities from the Try This! section located on pages 34–37 of the *Per Sense* Student Book. These activities can be used to reinforce the main concepts of Section C.

Planning Assessment

• Problem 8 can be used to informally assess students' ability to recognize the relationship between a fraction, a ratio, and a percent; to construct and use a visual model of percent; to determine which strategy for finding the percent is most appropriate in a given situation; and to construct and use a ratio table to find what percent is equivalent to a given fraction or ratio, or vice versa.

• Problems 12 can be used to informally assess students' understanding of the relative nature of percents and their ability to determine whether or not percents are used appropriately in a decision-making situation and to understand that fractions, ratios, and percents are used as comparison tools.

• Problem 24 can be used to informally assess students' ability to estimate percents and to determine which strategy for finding the percent is most appropriate in a given situation.

• Problem 26 can be used to informally assess students' ability to determine whether or not percents are used appropriately in a decision-making situation and their understanding of the fact that fractions, ratios, and percents are used as comparison tools.

• Problem 27 can be used to informally assess students' ability to determine which strategy for finding the percent is most appropriate in a given situation.

C. BENCHMARK PERCENTS

1. Do you have any idea how many people can be seated in the big stadium pictured above? Suppose the area that the arrow is pointing to contains 400 spectators. Estimate the total number of people in the stadium.

2. When it looks like it will rain, the number of fans drops by about 25 percent. About how many spectators will be in the stadium if it looks like rain?

1. Answers will vary. There are around 150 sections, and each section contains about 400 spectators. This makes a total of about 60,000 people.

2. Answers will vary. If students estimate about 60,000 people for problem 1, the answer to this question is about 45,000 people.

Overview Students estimate the number of fans in Dodger Stadium as shown in a photograph. They also estimate how many people are at a "rainy day" game, when the attendance is reduced by 25 percent.

About the Mathematics In this section, the different percent strategies are studied in greater depth. Students learn to recognize benchmark percents and develop strategies to easily find recognizable percents. They also estimate a more difficult quantity using benchmark percents.

Planning You may use problem **1** to briefly introduce the context of this section. Students may work in small groups on problems **1** and **2.** Be sure to discuss problem **1** with the whole class.

Comments about the Problems

1. It is very difficult to count each section one by one. Encourage students to use creative strategies for estimating the total number of spectators. For example, students might count the number of sections in the stadium and multiply that by the number of people in one section.

2. Many strategies are possible. Recall the strategies students used in Section A, such as "25 percent means the same thing as one quarter $(\frac{1}{4})$ of a given quantity."

Did You Know? The game of baseball most likely developed from an 18th-century English game called rounders. The two games differed in that in rounders, runners were actually hit with the ball, rather than tagged as in baseball. The modern game of baseball became popular in the U.S. with the troops during the Civil War. After the war, in 1871, the National Association of Professional Baseball Players was formed.

Source: Encyclopædia Britannica

The San Francisco Giants are playing the Los Angeles Dodgers. Dodger Stadium is filled to capacity. The Giants are winning, but sadly, their fans are in the minority.

3. How many Giants fans and how many Dodgers fans might be there? (Hint: Use your estimate of the total number of fans in the stadium from problem **1.**)

4. Describe the number of fans for each team in another way.

There are different ways to do this. For instance, you could make a bar in your notebook and shade in the parts of the bar representing the Dodgers fans and the Giants fans.

You could also use fractions:
　?_ Dodgers fans and _?_ Giants fans

Or you could use percents:
　?_ % Dodgers fans and _?_ % Giants fans

3. Answers will vary, depending on students' estimate for problem **1.** The number of Giants fans should be any number less than half of the total capacity, and the number of Dodgers fans must make up the remaining proportion. Accept all answers that meet these conditions.

4. Answers will vary. As an example, use 50,000 Dodgers fans and 10,000 Giants fans. The strategy here is to find half of the bar (30,000), then divide the halves into thirds (10,000, 20,000, 30,000 and 40,000, 50,000, 60,000). Benchmark percents of 25 percent and 75 percent could be used as reference points to estimate 20 percent and 80 percent.

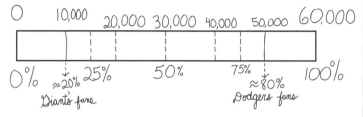

In this example, the percents would be as follows:
Giants Fans: about 20 percent
Dodgers Fans: about 80 percent

The fractions would be:

Giants Fans: about $\frac{10,000}{60,000}$ or $\frac{1}{6}$

Dodgers Fans: about $\frac{50,000}{60,000}$ or $\frac{5}{6}$

Overview Students estimate the numbers of Giants and Dodgers fans at one game. They then describe these numbers using fractions and percents.

About the Mathematics When the absolute numbers of a part-whole relationship are known, a percent bar is a strong visual model with which to estimate or calculate the percent that expresses the relationship. For example, here is how a percent bar can be used to express the ratio 300 out of 2,000 as a percent.

First, use benchmark percents to divide the blank percent bar into equal subparts. Write the absolute numbers that correspond to each benchmark percent as shown below.

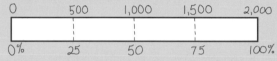

To get a quick estimate, use this strategy: estimate the location of 300 on the bar using the benchmark percents as reference points.

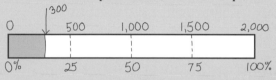

To get a more precise estimate, use this strategy: subdivide the indicated section using benchmark percents again, as shown.

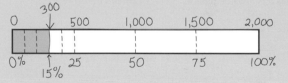

Planning Students can continue working on problems **3** and **4** in small groups. Discuss the strategies used in problem **4** with the whole class.

Comments about the Problems

3. Students should still be using their previous estimate of the total capacity of Dodger Stadium.

4. Precise calculations are not necessary. The percents can be found by estimating or by using a percent bar.

Suppose that there are 2 Giants fans for every 23 Dodgers fans.

5. a. Make a table like the one shown below. Find how many Giants fans there would be for 100 spectators.

   b. Make a percent bar to express the number of Giants fans as a percent of the total.

Giants Fans	2			
Dodgers Fans	23			
Total	25			

6. In problem **5,** why do you think you were asked to find the number of Giants fans out of 100?

Dodger Stadium holds about 60,000 people.

7. If the stadium is full, how many fans will each team have if there are 2 Giants fans for every 23 Dodgers fans? Express your answer using a percent bar.

8. Suppose there are 3 Giants fans for every 17 Dodgers fans. What will the number of fans for each team be then? Express the numbers of Giants fans and Dodgers fans as percents.

5. a. Ratio tables will vary. There would be 8 Giants fans for 100 spectators.

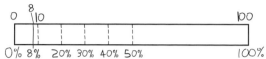

Giants Fans	2	4	8
Dodgers Fans	23	46	92
Total	25	50	100

b. Using the information from problem **5a,** students should be able use a percent bar to find that the Giants fans make up about 8 percent of the spectators.

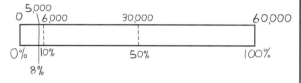

6. Since percent is a ratio of a number out of 100, if you find that there are 8 Giants fans out of 100 spectators, you know that this ratio is equal to 8 percent.

7. 4,800 Giants fans and 55,200 Dodger fans. From problem **5b,** students know that 2 is 8 percent of 25. Several strategies are possible using a percent bar. Students may find 10 percent on the percent bar and then subtract a little to estimate the total. Others might find 10 percent of 60,000 (6,000), find 2 percent (1 percent = 600, 2 percent = 1,200), and then realize that 10 percent − 2 percent = 8 percent. (Thus, 6,000 − 1,200 = 4,800 Giants fans, the precise answer. The number of Dodgers fans would then be 60,000 − 4,800, or 55,200 fans.)

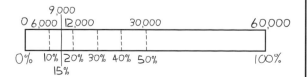

8. For 60,000 total spectators, there would be 9,000 Giants fans and 51,000 Dodgers fans. This equals 15 percent Giants fans and 85 percent Dodgers fans.

Overview From given ratios of Giants fans to Dodgers fans, students develop different strategies to find the percents and the numbers of fans for each team.

About the Mathematics The ratio table is another model with which to scale quantities to parts per hundred. For example, suppose the ratio of Giants fans to Dodgers fans is 3 to 17. A ratio table can be used to find out how many Giants fans there would be per 100 spectators.

Giants Fans	3	6	12	15
Dodgers Fans	17	34	68	85
Total	20	40	80	100

By systematically building up a given ratio to a ratio that is scaled by 100, students gain an understanding of equivalent fractions; of the concept that ratios are dependent upon the relationship between numbers, not the size of the numbers themselves; and of the fact that percents are just special cases of each.

Planning Students may work in pairs or in small groups on problems **5–7.** Then discuss students' answers and strategies for problems **6** and **7** with the whole class. You may ask students to work individually on problem **8.**

Comments about the Problems

5. Notice that the total number is a factor of 100. This should help bridge the gap to percents. For problem **5b**, estimations may vary, depending on the precision of students' strategies.

6. If students are having difficulty, ask them: *How many Giants fans are there in 100 total fans?* [8] and then work toward the concept of percents as fractions with denominators of 100.

7. It is important for students to realize that 2 Giants fans for every 23 Dodgers fans means 2 Giants fans out of every 25 visitors.

8. Informal Assessment This problem can be used to informally assess students' ability to construct and use a visual model of percent, to determine which strategy for finding the percent is most appropriate in a given situation, and to construct and use a ratio table to find what percent is equivalent to a given fraction or ratio, or vice versa.

Again, the numbers here divide nicely into 100 to help bridge the gap between fractions and percents. If students are having difficulty, you may suggest that they build a ratio table and use it to draw a percent bar.

BASEBALL SOUVENIRS

Finding out what team a person likes is easy. Just look at what he or she is wearing.

Before the ball game, a group of fans were asked about their favorite baseball souvenirs. They could choose any one of three items: a cap, a scarf, or a patch. Below you see their answers.

Results of Interview	Number of Fans Interviewed	Number of Fans Choosing a Cap	Number of Fans Choosing a Scarf	Number of Fans Choosing a Patch
Giants	310	123	89	91
Dodgers	198	119	20	58

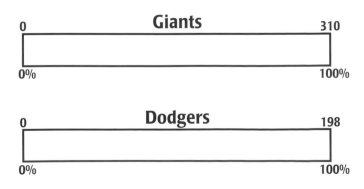

9. What do these statistics tell you?

10. Which group of fans, Dodgers or Giants, likes caps more? Estimate by making percent bars like the ones you see here.

11. Do the same for the scarf and patch. Which fans like the scarf more? Which fans like the patch more?

9. Both groups of fans like caps best, then patches, then scarves. But the percents are different for each team's fans.

10. Dodgers fans like caps more (60 percent of Dodgers fans compared to 40 percent of Giants fans).

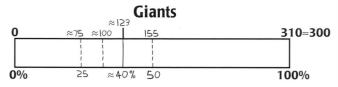

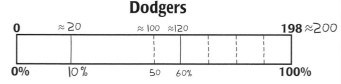

11. Percents for scarves are about 30 percent for the Giants fans and 10 percent for the Dodgers fans. Percents for the patches should be close to 30 percent for fans of each team. The bars below illustrate a sample strategy.

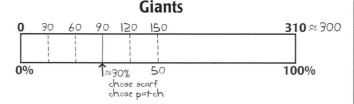

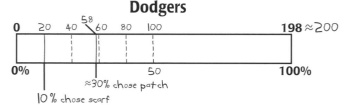

Overview Using the results of a survey, students compare the popularity of different souvenirs among the Dodgers and the Giants fans. They use percent bars to estimate the percents needed to make the comparisons.

About the Mathematics Percent bars can be used to compare different parts from different samples. Both the percent bars and the derived percents standardize the different part-whole relationships. The context of these problems is related to problems in the statistics strand. Simulations about population proportions, measured in percents, are studied more extensively in the grade 6/7 unit *Dealing with Data*.

Planning Students may work in pairs or in small groups on problems **9–11**. Discuss these problems with the whole class and let students share their answers and strategies.

Comments about the Problems

9. Let students discuss the data using number totals and percents to help them draw conclusions about the fans' preferences. Remind students that in order to make fair comparisons, they need to take the total number of fans into account. They should understand that a direct comparison can be made only if the total number of fans in each group is the same.

10. Some students will discover that the 10 percent benchmark is useful in making their estimations. Point out to students that although the number of Giants fans who chose the cap is larger than the number of Dodgers fans who chose a cap, the percent is smaller: 40 percent compared to 60 percent. This is yet another example of the relative nature of percent.

11. After students examine these percents, they can draw some conclusions about the different groups. Those familiar with the climate of California may think that San Francisco, being colder than Los Angeles, might have more fans preferring scarves.

THE SHOUTING MATCH

Two high school teams, the Camden Tigers and the Marquis Thunderbirds, were competing in a softball game. After the game, the coach of the Tigers said, "We had a greater percentage of fans in the stands this time compared to the last game."

"No, I don't think so," replied the pitcher, Julia. "I think they just made more noise this time because we were winning."

The table below shows the actual attendance figures.

12. a. Decide whether the coach or Julia was correct. You might draw a percent bar to help you find the answer.

	Total Number of Spectators	Number of Tigers Fans
Present Game	1,600	512
Previous Game	1,350	459

b. Describe any difficulties you might have had solving problem **12a.**

12. a. Both games attracted about the same percent of Tigers fans: around 33 percent. Students might use percent bars like these:

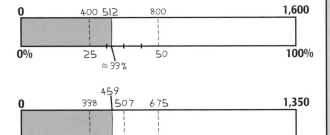

or like these:

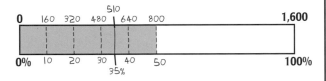

b. Answers will vary. Some students might have had difficulty constructing percent bars using such large numbers. Other students might have found it difficult to estimate where 512 falls between 400 and 800 in the first percent bar. Others might have found it hard to subdivide the second percent bar correctly to estimate where 459 should be located.

Overview Students compare the percents of Tigers fans at two different games using the given numbers in a table.

About the Mathematics Some students may implicitly use the "one percent strategy" in which they take one percent of the total and multiply that number until a reasonable approximation of the number of Tigers fans is reached. Do not teach this strategy, however. It will be formalized in the Summary as the one percent benchmark.

Planning Students may work on problem **12** in pairs or in small groups.

Comments about the Problems

12. Informal Assessment This problem can be used to assess students' understanding of the relative nature of percents, their ability to determine whether or not percents are used appropriately in a decision-making situation, and their ability to construct and use a visual model of percent. It also assesses whether or not they understand that fractions, ratios, and percents are used as comparison tools.

If students use estimation strategies, it is very difficult to find out if there are any differences in the two proportions. See if any students who come up with new strategies for finding exact percents on their own.

More Tiger Tales

Several years ago, the Tigers had a very bad season. They lost so many games that their fans stayed away. The Tigers will never forget one particular game. Below is a newspaper article about the game.

13. Estimate the percent of people at the game who were Tigers fans.

TIGERS SUPPORT WANES

TUESDAY'S GAME ATTENDED BY 1,598 — ONLY 16 TIGERS FANS

Tuesday's softball game was a low point in the Tigers fans' history.

Game	Total Number of Spectators	Tigers Fans
Tigers vs Thunderbirds	1,408	141
Tigers vs Wildcats	1,598	16
Tigers vs Grizzlies	1,212	606
Tigers vs Vikings	1,588	397
Tigers vs Moths	1,525	549
Tigers vs Cardinals	1,633	653
Tigers vs Lancers	987	187
Tigers vs Ground Squirrels	1,600	512
Tigers vs Huskies	1,350	459

On the left is the attendance record for the last nine games.

14. Rank the games in order, from the one with the smallest to the one with the largest number of Tigers fans.

15. Now rank the games in order from the smallest to the largest *percent* of Tigers fans.

16. Explain why there is a difference between the two lists.

13. Sixteen is about 1 percent of 1,598.

14. Rank order from smallest to largest number of Tigers fans:

Tigers vs Wildcats	16
Tigers vs Thunderbirds	141
Tigers vs Lancers	187
Tigers vs Vikings	397
Tigers vs Huskies	459
Tigers vs Ground Squirrels	512
Tigers vs Moths	549
Tigers vs Grizzlies	606
Tigers vs Cardinals	653

15. Rank order from smallest to largest percent (rounded to the nearest whole percent) of Tigers fans:

Tigers vs Wildcats	≈1 percent
Tigers vs Thunderbirds	≈10 percent
Tigers vs Lancers	≈19 percent
Tigers vs Vikings	25 percent
Tigers vs Ground Squirrels	32 percent
Tigers vs Huskies	34 percent
Tigers vs Moths	36 percent
Tigers vs Cardinals	≈40 percent
Tigers vs Grizzlies	50 percent

16. There is a difference in the rank orders because the rank order by percents is based on the ratio of the number of Tigers fans to the total number of fans in attendance, while the rank order by number only takes into account the number of Tigers fans at each game. In ranking by percent, a large number of Tigers fans compared to a small number of total fans (the Tigers vs Grizzlies game for example) might generate a larger percent than a large number of Tigers fans compared to a large total number of fans (the Tigers vs Cardinals game for example).

Overview Students rank the numbers of Tiger fans in different games in two ways. First, by looking at the actual numbers of Tiger fans and then by looking at the numbers of Tiger fans in relation to the total numbers of spectators at different games. Students make percent estimates for these comparisons.

About the Mathematics Data are compared *absolutely* and *relatively*. These kinds of comparisons often return in various units. These concepts are made explicit in the grade 6/7 unit *Ratios and Rates.*

Planning Students may work in small groups on problems **13–16.** It is advisable not to let students use calculators for these problems. Discuss problem **15** with the whole class, letting students explain how they recognized the percents.

Comments about the Problems

13. The numbers are chosen in such a way that the students should be able to see that 16 is about one percent of 1,598.

15. If students are having difficulty, ask them to try to identify which games have numbers that may be easy to express as percents. This may vary from student to student. (The numbers in the Tigers vs Grizzlies game can be easily expressed as 50 percent. The numbers in the Tigers vs Wildcats game can be easily expressed as about one percent.)

To estimate the percents for the last two games in the table, students should refer to problem **12** on page 21 of the Student Book, which uses the same numbers. The percent for the Tigers vs Moths game can be figured out by estimating with the fraction bar or by computing using a known percent.

16. Try to steer the discussion toward the relative nature of percent.

Getting In—*Getting Out*

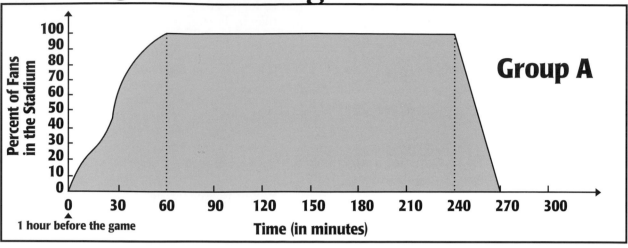

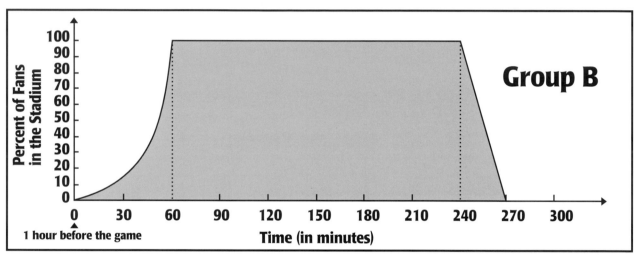

Recall the problems about Dodger Stadium (pages 17–19). The two **graphs** above show the percents of the two groups of fans in the stadium before, during, and after the game.

17. Use the graphs to answer the following questions.
   a. How long did it take all the people to get into the stadium?
   b. How long did it take them to get out?
   c. How long was the game?
   d. Was the score one-sided or close all the way to the end?

18. Which is the graph of Dodgers fans? Which is the graph of Giants fans? Why do you think so?

19. Remember in Dodger Stadium there were 2 Giants fans for every 23 Dodgers fans, so the number of fans differed greatly. Nevertheless, the two graphs look very similar. Why?

 Britannica Mathematics System

17. a. It took about one hour for all of the fans to file into the stadium.

b. It took about a half hour to file out of the stadium.

c. The game lasted approximately three hours.

d. Answers may vary. Since the fans for both teams stayed for the entire game, it was probably a close game.

18. Answers may vary, but should relate logically to the graphs. Examples of student answers follow:

Group A must be Dodgers fans. Since there are so many fans, they had to come early.

Group A must be Giants fans. Since there are so few Giants fans, 60 percent of them could easily get into the stadium during the last 30 minutes before the game.

Group B must be Dodgers fans because there are so many fans that they have to wait and are crowded at the entrance gates.

Group B must be Giants fans. Since there are so few Giants fans, 90 percent of them could easily get into the stadium during the last 30 minutes before the game.

19. The two graphs are similar because they are standardized in terms of percents. The maximum percent of fans in the stadium for the two teams is 100 percent. However, the maximum number of fans at a ratio of 2 to 23 is 4,800 Giants fans to 55,200 Dodgers fans. Although the numbers differ greatly, this difference will not show up on the graphs.

Overview Students read and interpret information from two graphs that show the percent of fans in the stadium during different time periods.

About the Mathematics Reading data from graphs is one of the topics in the grade 5/6 unit *Picturing Numbers.* If students have not yet studied this unit, some may have difficulties reading graphs. If students have already completed the unit *Picturing Numbers,* they should be able to find the appropriate time on the horizontal axis and read off the appropriate percent. (See the graph below.)

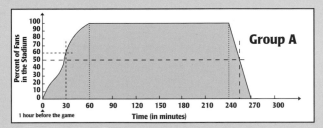

Planning Students may work in pairs or in small groups on these problems.

Comments about the Problems

17. A key question with which to spark discussion might focus on the difference in the first hour for both graphs. What does this difference represent in real terms? The fans of Group A filed into the stadium fairly steadily: after 30 minutes, 60 percent of the group had entered the stadium. Most of the fans of Group B went in late: after 30 minutes, only about 10 percent of them had entered the stadium.

18. Student response will depend on the assumptions they have made about how home team fans and visiting team fans get in and out of stadiums. If they think that visiting fans rush in at the last minute, while home team fans arrive early, then the top graph might be that of the Dodgers fans and the bottom graph that of the Giants fans. This problem is a good example to illustrate that different answers are possible, as long as the reasoning is correct.

19. The two graphs are similar because they are standardized in terms of percents.

	Total Number of Spectators	Number of Dodgers Fans	Number of Giants Fans
Half an Hour before the Game Started			
During the Game	60,000	55,200	4,800
Fifteen Minutes after the Game Ended			
One Half Hour after the Game Ended			

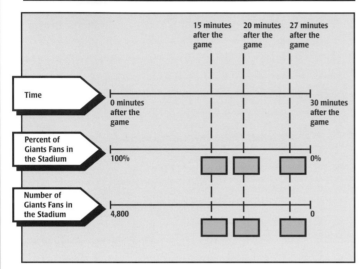

20. Take another look at the two graphs on page 23. Assume that Group A is the Dodgers. Fill in the table on **Student Activity Sheet 9** with the appropriate number of fans for each category. Use the space at the bottom of the activity sheet to explain how you got your answers.

21. Complete the diagram on **Student Activity Sheet 10.** Fill in the boxes for the percent and the number of Giants fans in the stadium 15, 20, and 27 minutes after the game.

22. Suppose the Giants were so far behind that they had no chance of winning. What would the "getting in—getting out" graph for the Giants fans look like then?

20. See the table below.

	Total Number of Spectators	Number of Dodgers Fans	Number of Giants Fans
Half an Hour Before the Game Started	≈ 33,700	33,000	≈ 700
During the Game	**60,000**	**55,200**	**4,800**
Fifteen Minutes After the Game Ended	30,000	27,600	2,400
One Half Hour After the Game Ended	0	0	0

21. Students' strategies may vary. Students may use benchmark percents or percent bars to obtain their answers. See the diagram below.

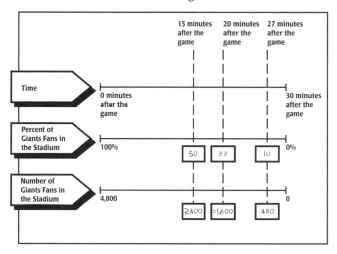

22. Answers will vary. The part of the graph that shows the Giants fans getting out of the stadium might start earlier and/or might be curved. One could argue that the Giants fans would start to trickle out when the Giants started losing, and the number of fans leaving would increase as it became clear that the Giants could not win.

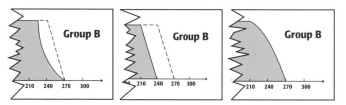

Materials Student Activity Sheets 9 and 10 (one of each per student)

Overview Students use the graphs to find the percents of Dodgers and Giants fans for specific time periods. They then use these percents to find the actual numbers of Dodgers and Giants fans in the stadium at these times.

About the Mathematics In problem **22,** students describe a part of the graph globally. The idea of describing or looking globally at the shape of the graph is one of the topics in the grade 6/7 unit *Tracking Graphs.* The concepts of increase and decrease will be treated more extensively in the grade 7/8 unit *Ups and Downs.*

Planning Students may work in small groups on problems **20–22.**

Comments about the Problems

20. Students should be able to find the indicated time period on the horizontal axis and read the appropriate percent from the vertical axis. Then students use the percents to find the numbers of Giants and Dodgers fans.

21. If students simply read the percents from the vertical axes of the graphs, they may not get very accurate answers for the percents of fans in the stadium at 20 minutes after the game and 27 minutes after the game.

Some student may solve this problem using ratios: 100 percent of the fans leave in 30 minutes, so 10 percent of the fans leave every three minutes.

22. If students are having difficulty visualizing the possible changes in the Group B graph, draw the three examples (shown below on the left) on the board. Ask students which graph they would choose and why.

The SCHOOL TRIP

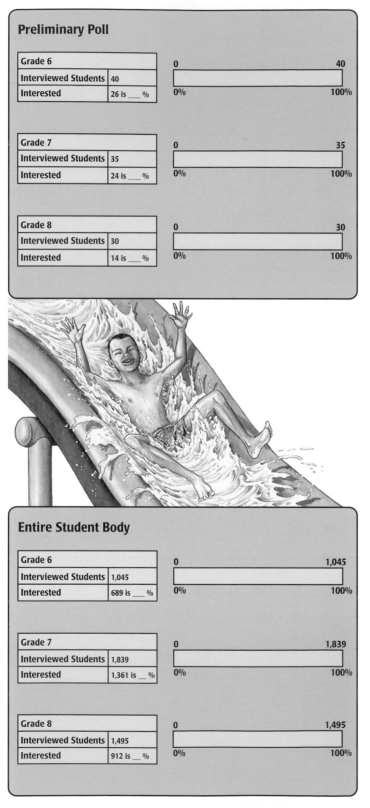

Preliminary Poll

Grade 6		0	40
Interviewed Students	40		
Interested	26 is ___ %	0%	100%

Grade 7		0	35
Interviewed Students	35		
Interested	24 is ___ %	0%	100%

Grade 8		0	30
Interviewed Students	30		
Interested	14 is ___ %	0%	100%

Entire Student Body

Grade 6		0	1,045
Interviewed Students	1,045		
Interested	689 is ___ %	0%	100%

Grade 7		0	1,839
Interviewed Students	1,839		
Interested	1,361 is ___ %	0%	100%

Grade 8		0	1,495
Interviewed Students	1,495		
Interested	912 is ___ %	0%	100%

Students at Martin Luther King Middle School are planning an end-of-the-year field trip. The student council has suggested going to Noah's Ark water park. Before making a final decision, they want to find out how many students are interested in going.

The student council decided to **poll** selected classes in grades 6, 7, and 8 to find out the percent of the student body in favor of going to Noah's Ark. The results of their poll are shown on the left.

23. **a.** Fill in the percent bars on **Student Activity Sheet 11.**

   b. What is the percent of students for each grade level who are interested in going to Noah's Ark?

The results of the preliminary poll were not as the student council expected, so they decided to poll the entire student body.

24. **a.** Fill in the second set of percent bars on **Student Activity Sheet 11.**

   b. What is the percent of students for each grade level who favor going to Noah's Ark now?

25. Do you think the preliminary poll was an accurate assessment of students' interest? Why or why not?

26. Based upon the results of the student body **survey,** would you recommend going to Noah's Ark for the school trip? Explain your reasoning.

23. a. Accept percent estimates close to these answers.

Preliminary Poll

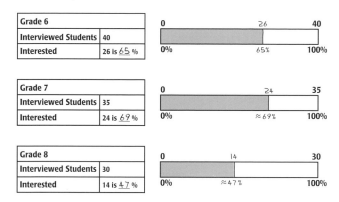

Grade 6	
Interviewed Students	40
Interested	26 is <u>65</u> %

Grade 7	
Interviewed Students	35
Interested	24 is <u>69</u> %

Grade 8	
Interviewed Students	30
Interested	14 is <u>47</u> %

b. Grade 6: about 65 percent
Grade 7: about 69 percent
Grade 8: about 47 percent

24. a. Accept percent estimates close to these answers.

Entire Student Body

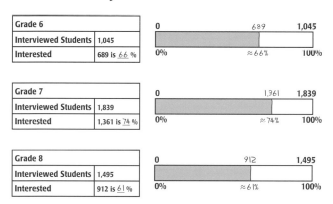

Grade 6	
Interviewed Students	1,045
Interested	689 is <u>66</u> %

Grade 7	
Interviewed Students	1,839
Interested	1,361 is <u>74</u> %

Grade 8	
Interviewed Students	1,495
Interested	912 is <u>61</u> %

b. Grade 6: about 66 percent
Grade 7: about 74 percent
Grade 8: about 61 percent

25. Answers will vary. Some students might say that the preliminary poll was fairly accurate since the percent estimates for Grades 6 and 7 are about the same in both polls. Other students might say that the preliminary poll was not accurate since the percent estimates for Grade 8 are different in both polls.

26. Answers will vary. The majority of students (68 percent) favor going to Noah's Ark. This is slightly more than $\frac{2}{3}$ of the student body. Students may feel that this is enough of a majority, or they may feel that there should be overwhelming support in order to choose the water park.

Materials Student Activity Sheet 11 (one per student)

Overview Students interpret the results of two polls using percent to determine the portion of each grade level who voted in favor of going to Noah's Ark water park.

Planning Students may work in small groups on problems **23–26**. These optional problems are ideal for homework, assessment, or additional practice. You may elect to skip these problems if your students are progressing nicely, or if time is a concern.

Comments about the Problems

23. Homework This problem may be assigned as homework. Students may use any strategy to find the answers. The percent bars are included as aids for students who might need them.

24. Informal Assessment This problem assesses students' ability to estimate percents and their ability to determine which strategy for finding the percent is most appropriate in a given situation. Any strategy is acceptable.

25. Homework This problem may be assigned as homework. Although the percents for each grade level are different in each poll, it is unclear whether these slight differences are significant. Let students struggle with this concept that is very important in the study of statistics.

26. Informal Assessment This problem assesses students' ability to determine whether or not percents are used appropriately in a decision-making situation and their understanding of the fact that fractions, ratios, and percents are used as comparison tools.

The important issue is this: Can students use percents to defend their arguments?

Summary

In this section, you saw that certain percents are easier to express and work with than others. Some of these **benchmark percents** are 1%, 10%, 25%, 33%, and 50%.

Summary Question

Each year, the Chamber of Commerce of New Lincoln holds a triathlon to raise money for local charities. Because the race is so grueling, a number of people drop out each year. Tonya used the following strategies to find the percent of dropouts for each of the five years listed below.

Year of Triathlon	Number of Competitors	Number of Dropouts	Percent of Dropouts	Tonya's Strategy
1988	1,493	15	about 1%	I saw it immediately. The number 1,493 is about 1,500; 15 is 1% of 1,500. So about 1% dropped out.
1989	2,890	1,445	50%	The number of dropouts,1,445, is half of 2,890, so 50% dropped out.
1990	1,194	401	about 33%	When I rounded off the numbers to 1,200 and 400, I discovered that 401 is about one-third of 1,194. So about 33% dropped out.
1991	1,550	310	20%	I thought that 310 times 5 is 1,550, so 20% dropped out.
1992	798	323	about 40%	If 1% dropped out, there would be about 8 dropouts. Since there were 323 dropouts, 323 divided by 8 makes about 40%.

27. Now it is your turn. Use **Student Activity Sheet 12** to find the percent of dropouts from the marathon portion of the triathlon. Describe your strategies.

27. Answers will vary. See sample chart below.

Year of Marathon	Total Number of Competitors	Number of Dropouts	Percent of Dropouts	Describe Your Strategy
1988	1,340	670	50%	Benchmark
1989	1,621	392	≈25%	Benchmark, Rounding
1990	1,793	180	≈10%	Round 1793 up to 1800, 180 is $\frac{1}{10}$ of 1,800, so it is equal to 10%
1991	1,603	91	5%–6%	1% of 1,600 = 16 16 × 5 = 80 16 × 6 = 96 Between 5% and 6%
1992	1,400	350	25%	4 × 350 = 1400 So 350 is $\frac{1}{4}$ of 1,400, or 25%

Materials Student Activity Sheet 12 (one per student)

Overview Students read in the Summary that the "easy percents" they have encountered are known as *benchmark percents*. They also read and discuss examples of the various strategies Tonya used to solve percent problems. Students also solve percent problems using similar strategies.

Planning After students have read the Summary, discuss the strategies Tonya used to solve the problems on this page. Strategies include using the one percent strategy and finding benchmark fractions. After the discussion, have students work individually on problem **27**. After students complete Section C, you may assign appropriate activities from the Try This! section, located on pages 34–37 of the Student Book, as homework.

Comments about the Problems

27. Informal Assessment This problem assesses students' ability to estimate percents and to determine which strategy for finding the percent is most appropriate in a given situation.

Encourage students to describe their strategies and to find more than one way to solve each percent problem. Ask students which year's triathlon was the most grueling. Have them explain their reasoning.

SECTION D. A FINAL TIP

Work Students Do

Students decide how much of a tip to leave for a lunch bill of $4 and determine what percent of the bill their tip represents. Students then find the tips for different lunch bills using the same percent as was left for the $4 lunch. To decide which meal the Spanish club will order at a restaurant, students compute and compare the total costs of four different meals, including tip, for 25 students. They solve additional problems in which they estimate what percent of a bill is represented by a given tip. Students then determine 10 and 15 percent tips for different bills.

Finally, students use their knowledge and understanding of percents to solve an economic crime.

Goals

Students will:

- estimate percents;
- use benchmark percents of 1%, 10%, 25%, and 50%;
- construct and use a ratio table to find what percent is equivalent to a given fraction or ratio, or vice versa;
- construct and use a visual model of percent;
- recognize the relationship between a fraction, a ratio, and a percent;
- understand the relative nature of percent;
- determine whether or not percents are used appropriately in a decision-making situation.

Pacing

- approximately four 45-minute class sessions

About the Mathematics

In this section, the key concepts and strategies of the unit are repeated and extended. The percent strategies include estimating percents and calculating exact percents using benchmark fractions such as $\frac{1}{4}$ and $\frac{1}{2}$, benchmark percents such as one percent and ten percent, percent bars, and ratio tables.

The context of tipping in a restaurant is related to the mathematical topic of growth situations.

Planning Instruction

The problems in this section are no more difficult than those in Sections B and C. Because this is the final section of the unit, it is important that students are able to do most of the problems without help. When they have succeeded in doing these problems, they will have a positive feeling about what they have learned in the unit. After students complete the problems, let them discuss their strategies and reasoning with the whole class.

Students may work in pairs or in small groups on problems 1–7 and 15–20. You might want students to work individually or in pairs on problems 8–14 and 19–22.

There are no optional problems. Be sure to discuss problems 1, 6–10, 12, and 13 with the whole class.

Homework

Problems 10–11 (page 68 of the Teacher Guide), problems 21–22 (page 76 of the Teacher Guide), and the Writing Opportunity (page 75 of the Teacher Guide) may be assigned as homework. After students complete Section D, you may assign appropriate activities from the Try This! section located on pages 34–37 of the *Per Sense* Student Book. These activities can be used to reinforce the main concepts of Section D.

Planning Assessment

- Problem 3 can be used to informally assess students' ability to estimate percents.

- Problems 12 and 14 can be used to informally assess students' ability to estimate percents; to construct and use a ratio table to find what percent is equivalent to a given fraction or ratio, or vice versa; to use benchmark percents; to construct and use a visual model of percent; and to recognize the relationship between a fraction, a ratio, and a percent.

- Problem 15 can be used to informally assess students' understanding of the relative nature of percent.

- Problem 19 can be used to informally assess students' ability to determine whether or not percents are used appropriately in decision-making situations, their understanding of the relative nature of percent, and of the fact that fractions, ratios, and percents are used as comparison tools.

D. A FINAL TIP

Don't Forget the Tip

1. If the bill for your lunch was $4.00, what would you leave as a tip? In your notebook, draw the coins that you would leave on the table.

2. What percent of the $4.00 total did you give as a tip? Make a percent bar in your notebook to figure it out.

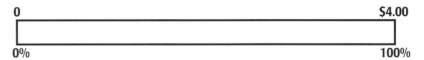

0          $4.00

0%   100%

3. Find the percent in another way. Explain how you did it.

4. If you continued to tip as you did on the $4.00 bill, what tip would you leave on bills of other amounts? Copy and complete the table below. Then extend it with two more examples.

Bill	$4.00	$2.00	$1.00	$5.00	$7.00		
Tip							

5. Make another table like the one above. Write the money amounts in cents rather than in dollars.

6. Look at the gray column above. How could you use the money amounts in this column to find a percent?

         Britannica Mathematics System

1. Answers will vary.

2. Answers will vary, depending on students' responses to problem 1. If the tip amount is not an easy fraction of $4.00, accept a percent estimate.

3. Strategies will vary. A ratio table is one possibility. Students may also simply subdivide the percent bar differently.

4. Answers will vary depending on students' responses to problem 1.

   Sample ratio table:

Bill	$4.00	$2.00	$1.00	$5.00	$7.00	$10.00	$20.00
Tip	60¢	30¢	15¢	75¢	1.05¢	1.50¢	3.00¢

5. This will be the same ratio table as above, but with the dollar amounts written as cents.

6. Since $1.00 consists of 100 cents, any division of it becomes a percent. The tip for any multiple of $1.00 is the base percent times the multiplying factor (for example, if the tip amount for $1.00 is $0.40, then the tip amount for $4.00 is 4 × $0.40, or $1.60).

Overview Students decide how much money they would leave as a tip for a bill of $4.00. They then find out what percent of the $4.00 their tip represents.

About the Mathematics In this section, the key concepts presented in the first three sections are repeated and extended. This is especially true for the ratio table model and benchmark percents, such as 1 percent and 10 percent.

Planning Students may work in pairs or in small groups on problems **1–6.** Discuss the answers and strategies for problems **1** and **6** with the whole class.

Comments about the Problems

1. This problem gives you an opportunity to see what students know about tipping in restaurants. Discuss what tip they gave and why they gave that much (or that little). Be sure they are aware that the tip is added to the original meal price to determine the total amount they should pay.

2. The difficulty of this problem depends upon the tip students chose to leave in problem **1.** If the tip is not an "easy fraction" of $4, Encourage students to estimate the percent of the tip.

3. **Informal Assessment** This problem assesses students' ability to estimate percents.

5. This activity is designed to avoid the confusion caused sometimes when computing with decimals. All numbers are converted to whole numbers, using cents as the unit.

6. This question stresses that percent means "so many out of 100," and that percents can be used to standardize different quantities so they can be compared directly. If students understand these two concepts, then all the goals of the unit have been reached. Be aware that some students can have an understanding of a concept without being able to verbalize their understanding in a formal way.

MUCHO DINERO

$

The Spanish club is planning an end-of-the-year banquet at a local Mexican restaurant. The restaurant is willing to provide a private dining room and a choice of four different meals for the 25 students who will attend.

The only requirements are that everyone attending must order the same meal and a 15% tip will be added to the total bill.

The club treasurer needs to know the total cost of each type of meal, including the tip.

7. Copy and complete the table below. Find the total cost, including the tip, for each of the four meals.

Meal	Cost per Meal	× 25 Students	Tip	Total Cost
La Fiesta Grande	$5.00			
El Pollo Especial	$7.50			
El Combinacion Mexicano	$10.00			
El Cerdo Grande	$12.50			

7. See table below. Note: All tip amounts were rounded to the nearest cent.

Meal	Cost per Meal	× 25 Students	Tip	Total Cost
La Fiesta Grande	$5.00	$125.00	$18.75	$143.75
El Pollo Especial	$7.50	$187.50	$28.13	$215.63
El Combinacion Mexicano	$10.00	$250.00	$37.50	$287.50
El Cerdo Grande	$12.50	$312.50	$46.88	$359.38

Overview Students find the total cost, including a 15 percent tip, of four different meal choices for 25 students.

About the Mathematics The context of tipping in a restaurant is used to explore the strategy of finding a 10 percent tip, finding half of the 10 percent tip, and adding the two amounts to determine a 15 percent tip.

Planning Students may work in small groups on problem **7.** Discuss their strategies with the whole class.

Comments about the Problems

7. If students are having difficulty, suggest that they use a percent bar to find the tip amounts. Some students may use the idea of the previous problem: a 15 percent tip means 15 cents for every dollar. Other students may see a pattern in some of the meal costs in the table: the El Combinacion Mexicano is twice as expensive as La Fiesta Grande. If students do not see this pattern, have them compare these two meals prices in a class discussion.

Extension You may also suggest an alternate strategy to solve problem **7.** Ask students: *Suppose you first add the tip to the cost of one meal and then multiply that amount by 25. Will this make a difference in the total cost of the meals? Why or Why not?* [No. Since both steps involve a multiplication operation, the order of the steps is not important ($\$5 \times 25 \times \frac{15}{100} = \$5 \times \frac{15}{100} \times 25$).]

Monday Luncheon Special

On Monday, Kay ordered the daily special for lunch. Her bill came to $2.95, and when she finished, she left 35¢ as a tip on the table.

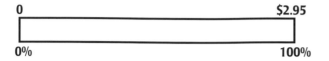

8. Estimate what percent of the bill Kay gave as a tip. Use a percent bar to help you.

35						
295						

9. Use a table to determine the percent Kay left as a tip. Try to change "35 out of 295" into "so many out of 100."

A third way to solve this problem is to use easily found percents. For instance, find 10% and 1% of 295. These percents can give you approximations for solving the problem.

10. Explain how to use 10% and 1% of 295 to find the percent Kay left as a tip.

Now you have learned three ways to calculate percents:

- estimating with a percent bar,
- changing the numbers into "so many out of 100" by means of a table,
- using easily found percents, such as 1% or 10%.

Where Does All the Money Go?

Most waiters and waitresses depend on tips for their income. Years ago, the standard tip for good service in a restaurant was 10% of the total bill. Now, restaurants are allowed to pay their staffs less than minimum wage, so the standard tip has risen to about 15% of the total bill. Of course, leaving a tip is optional, and customers often leave more or less than 15%, depending on the quality of the food and service.

11. What would 10% and 15% tips be for the following bills: $7.54, $9.78, $14.75, and $37.20? Explain your strategies.

8. Accept percent estimates close to 12 percent.

Sample percent bar:

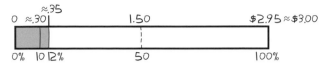

9. Answers will vary. Students may round 295 to 300.

Sample ratio table:

35	≈36	≈24	≈12
295	≈300	≈200	≈100

10. Answers will vary. Here are two sample explanations:

Ten percent of 295¢ is about 30¢, and 1 percent of 295¢ is about 3¢, so 35¢ is about 10 percent + 2 percent, or about 12 percent.

or

One percent of 295¢ is about 3¢; 36¢ is about 12 times 3¢, so 35¢ is about 12 × 1 percent, or about 12 percent.

11. Accept tip estimates close to these answers.

Bill	10% Tip	15% Tip
$ 7.54	$0.75	$1.13
$ 9.78	$0.98	$1.47
$14.75	$1.48	$2.22
$37.20	$3.72	$5.58

Note: All tip amounts were rounded to the nearest cent.

Strategies will vary. Sample strategies to find a 15 percent tip for a bill of $7.54:

Strategy 1

First I found what 10 percent of $7.54 is by multiplying the bill by one-tenth [$\frac{1}{10}$ × 7.54 ≈ $0.75].

Then I found what 5 percent of $7.54 is by dividing the 10 percent tip amount by two [0.75 ÷ 2 ≈ $0.38].

Then I added the 10 percent tip amount to the 5 percent tip amount [$0.75 + $0.38 = $1.13].

Strategy 2

First I found what 1 percent of $7.54 is by multiplying the bill by $\frac{1}{100}$ [$\frac{1}{100}$ × 7.54 ≈ $0.075].

Then I multiplied that amount by 15 to find 15 percent of the bill [15 × 0.075 ≈ $1.13].

Overview With a given amount and a given tip, students find what percent of the bill their tip represents using three different strategies. Students then calculate what 10 and 15 percent tips would be for four different bills.

About the Mathematics Different strategies are made explicit on this page. Students should now be able to use the percent bar, ratio table, and all the benchmark fractions and percents.

These strategies are directly related to many goals of this unit:

- estimate percent numerically using a percent bar,
- construct a ratio equivalent (some number out of 100) using a ratio table,
- find percents using the benchmark percents of 1 percent and 10 percent.

Planning You might want students to work individually or in pairs on problems **8–11** in order for you to evaluate their progress.

Comments about the Problems

8. This problem is similar to problem **1**, but now the tip amount is given. The given amount makes it necessary to use estimation. By rounding the bill amount to the nearest dollar, a close estimate is possible.

9. Discuss the differences in students' estimated answers. In the ratio table, the price is already converted to the same unit as the tip: cents.

Using the ratio table, many students will get a ratio of 12 cents out of 100 cents, or 12 percent. Some students may say that $\frac{12}{100}$ is close to $\frac{1}{10}$, or 10 percent.

10. Homework This problem can be assigned as homework. Some students may need a visual model, such as a percent bar, to find the percent.

11. Homework This problem can be assigned as homework. Any one of the percent strategies can be used here. Examine students' responses to see whether or not their strategies are becoming more sophisticated.

Ms. Watson's Royalties

Ms. Watson writes detective stories. Her most recent book, *The Case of the Missing Nose,* sells for $7.50. She gets 30 cents for every copy sold. This is called her *royalty.*

12. Use the three strategies on page 29 to find what percent of the selling price is Ms. Watson's royalty.

13. Which of the three strategies do you prefer? Why?

Ms. Watson likes to eat at Marleen's Cafe. Her bill today comes to $4.35. Ms. Watson decides to leave a "three-copies-tip," referring to the amount of royalties she receives from selling three copies of her book.

14. **a.** How much would the tip be?

b. What percent of the bill is the tip? Try to figure this out using each of the three strategies.

c. Which strategy do you prefer this time?

12. Strategies will vary. Here are four possible strategies:

Strategy 1

Thirty cents is a little less than 5 percent, or about 4 percent.

Strategy 2

30	15	75	3.75	≈4
750	375	1875	93.75	≈100

or

30	3	1	4
750	75	25	100

Strategy 3

Ten percent of $7.50 is $0.75; half of $0.75 (5 percent) is about $0.38, so $0.30 is a little less than 5 percent or about 4 percent.

Strategy 4

One percent of $7.50 is $0.075; 4 x $0.075 = $0.30, so $0.30 is 4 percent of $7.50.

13. Answers will vary. Students who do not like to draw might prefer using a ratio table to find a ratio with a denominator of 100. Students who are more visual might favor drawing a percent bar to estimate the percent. Still other students might be more comfortable using the 10 percent and 5 percent method or the 1 percent method because these use more of a step-by-step approach. Accept all answers that show logical reasoning for students' choice of strategies.

14. a. Three copies at 30¢ per copy is 90¢.

b. Ninety cents is about 21 percent of $4.35. Accept percent estimates close to 21 percent.

c. Answers will vary. See sample explanations for problem **13** above.

Overview Students again use the three strategies listed on the previous page to determine the percent of an author's royalty per book, given the royalty amount and the selling price of one book.

Planning Students may work individually or in pairs on problems **12–14**.

Comments about the Problems

12. The percents in this problem are smaller than those in the tipping problems. However, the same strategies can be used here.

14. This problem is similar to problem **12**, only the tip amount is not given directly. Students must read the text carefully to find the tip amount.

12, 14. Informal Assessment These problems assesses students' ability to estimate percents; construct and use a ratio table to find what percent is equivalent to a given fraction or ratio, or vice versa; use benchmark percents; construct and use a visual model of percent; and recognize the relationship between a fraction, a ratio, and a percent.

A special collectors' edition of Ms. Watson's first book has been published. For this more expensive book, her royalty per copy is 50 cents.

To celebrate this occasion, Ms. Watson decides to take her husband out for a fancy dinner. The bill comes to $57.85. Suppose Ms. Watson leaves a tip that is the same percent of the bill as the percent of her royalties.

15. How much money will she leave on the table? Would this be a reasonable tip? Why or why not?

Budget of Elbonia

International Aid	$6 billion
Military	$72 billion
Transportation	$12 billion
Social Security	$66 billion
Health	$12 billion
Agriculture	$6 billion
Natural Resources	$12 billion
Energy	$6 billion
Debt	$30 billion
Income Security	$45 billion
Education	$33 billion
Total	**$300 billion**

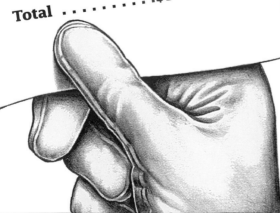

A Phony in Elbonia is a story about a crime involving money meant for the support of developing countries. It is about an incident in a fictitious country called Elbonia.

The budget for Elbonia is pictured on the left. Notice that $6 billion has been allocated for International Aid.

16. What percent of the total budget is this amount?

17. About $3 billion of the International Aid budget was missing at the end of the year. What percent was missing?

18. It is possible that your answer to problem **17** differs from other people's answers. See if everyone got the same answer. How is it possible for there to be different correct answers to this question?

15. There is not enough information to solve this problem. You need to know the book price in order to determine what percent 50¢ is of the book price.

16. Six billion dollars is about two percent of the total budget of $300 billion.

17. If $6 billion is about two percent, $3 billion must be about one percent of the total budget of $300 billion. Three billion dollars is also 50 percent of the International Aid budget of $6 billion.

18. Yes. Some students may have calculated their answer by finding what percent $3 billion is of the total budget. Other students may have calculated their answer by finding what percent $3 billion is of the International Aid budget.

Overview Students first encounter a word problem that has too little information to solve. They then encounter a problem that can be interpreted in two different ways to produce two different solutions.

About the Mathematics The focus of these problems is the relative nature of percent. Students need to ask themselves what each percent refers to as they solve each problem.

Planning Because of the difficult nature of these problems, students should not work on their own. Students may work in small groups on problems **15–18.** Let students share their strategies and opinions with each other.

Comments about the Problems

15. Informal Assessment This problem can be used to assess students' understanding of the relative nature of percents. This problem reiterates the fact that because of the relative nature of percents, percents can only be calculated in terms of some "whole."

16. Encourage students to calculate with these numbers using "billion" as a unit rather than writing the numbers in standard form, with nine zeros.

17–18. Problem **17** has two different solutions. The "whole" amount in this problem can refer to the total budget or to the International Aid budget.

You Be the Detective

The government of Elbonia is having problems accounting for all of the money spent.

Mr. Butler is the Elbonian bureaucrat whose job is to deliver the money to developing countries. For his work, he gets a 1% commission.

An undercover detective who is interviewing all the bureaucrats succeeds in getting a dinner appointment with Mr. Butler.

After dinner, the server brings the check to the table. The total is $20. Mr. Butler announces his intention to leave a 15% tip. First he gives the server a dollar. "That's 5%," he says. Then he adds a dime to the dollar. "This is another 10%, so altogether it is a 15% tip," he explains confidently.

The server is stunned and can't say a word. Mr. Butler looks at her with a smug expression. "You're welcome," he says. Suddenly, the detective jumps up and says "Aha! Now I know where the money went! You are under arrest!"

19. What did the detective figure out that could be used to convict Mr. Butler of fraud? In your notebook, write your answer as completely as possible so it can be used by the prosecuting attorney at Mr. Butler's trial. Include all the important information you know about percents so that the prosecuting attorney can convince the jury.

20. Is there any way that Mr. Butler could plan his defense? Explain.

19. Answers will vary. The fact is that Mr. Butler correctly calculated 5 percent of $20 to be $1. He then incorrectly calculated 10 percent of $20 by finding 10 percent of $1 to be $0.10, which is only 0.5 percent of $20, resulting in a 5.5 percent tip.

To calculate his commission, Mr. Butler could have used the same logic: He could have incorrectly taken 1 percent of the total budget (1 percent of $300 billion = $3 billion) rather than 1 percent of the International Aid budget (1 percent of $6 billion = $60 million).

20. Answers will vary. Here is a sample defense:

Mr. Butler may claim that you cannot make a comparison between the two isolated events by saying, "Just because I cheated the waitress on her tip doesn't mean that I cheated my country too!" He might also claim that he has never understood percents and has never been good with numbers. He just made an innocent mathematical mistake in calculating the tip and in calculating his commission.

Overview Students read a short detective story. They then use their knowledge and understanding of percents to explain why the detective in this story has accused Mr. Butler of stealing money from the International Aid budget.

A Synopsis of the Story In this crime story, Mr. Butler, a member of Elbonia's government, is accused of taking a $3 billion dollar commission, rather than the $60 million dollar commission to which he was entitled. He took 1 percent of the wrong "whole" amount—the total budget of Elbonia instead of the International Aid budget. The detective arrests Mr. Butler for the political crime after witnessing how Mr. Butler tried to cheat a server out of her tip at a restaurant. The detective suspects that Mr. Butler tried to use the same faulty mathematical reasoning in attempting to misdirect government funds into his own pocket.

Planning After reading the story together, students may work individually or in pairs on problems **19** and **20**. You may use problem **19** as an informal assessment, a writing opportunity, or a homework assignment.

Comments about the Problems

19. Informal Assessment This problem can be used to assess students' ability to determine whether or not percents are used appropriately in a situation of decision-making, their understanding of the relative nature of percent, and of the fact that fractions, ratios, and percents are used as comparison tools.

The logic in this problem may not be apparent to some students. Discuss how Mr. Butler knowingly cheated the waitress on her tip using faulty logic about percents. Then discuss the different possible bases from which the 10 percent could be calculated ($1 or $20). Ask students what percent tip Mr. Butler actually gave the waitress (5.5 percent).

Writing Opportunity Since student responses to the problems on this page may be rather lengthy, you might want to have students write their answers for problems **19** and **20** in their journals.

Summary

In this section, you calculated percents using three strategies:

- estimating with a percent bar,

- changing the numbers into "so many out of 100" by means of a table,

- using easily found percents, such as 1% or 10%.

Some of these strategies produce exact answers, and others produce approximate answers.

Summary Questions

21. Which strategies produce exact answers?

22. Write a percent problem in which an approximate answer is appropriate and another percent problem in which an exact answer is needed.

21. Answers will vary. You can sometimes get exact answers by using a ratio table or by using benchmark percents. Depending on the numbers involved in the problem, you can also get an exact answer using a percent bar.

22. Problems will vary. Sample problems follow.

Sample problem where an approximate answer is acceptable:

Our school auditorium can seat 528 people. If 319 people attended the spring musical review, what percent of the auditorium was occupied?

Sample problem where an exact answer is needed:

To get a passing grade in science this grading period, Jon needs to get a grade of 90 percent or better on his next test. On the next test, Jon got 8 problems correct out of the 10 test problems. What percent of the problems did he answer correctly on his science test?

Overview Students read and discuss the Summary, which highlights the three strategies used in this section to estimate or calculate percents. They then determine which of these strategies might be best for problems that require exact percent answers or percent estimate answers. Students also write two percent problems of their own: one problem for which an exact answer is needed and one problem for which an approximate answer is satisfactory.

Planning Students may work individually or in pairs on problems **21** and **22.** After students complete Section D, you may assign appropriate activities from the Try This! section, located on pages 34–37 of the Student Book, as homework.

Comments about the Problems

21–22. Homework Problems **21** and **22** can be assigned as homework. Problem **21** focuses on exact answers. Both the given numbers and the chosen strategy will influence the exactness of the final answer.

Assessment Overview

Students work on seven assessment problems that you can use to collect information about what each student knows about percents and what strategies they used to solve the problems.

Goals

- estimate percents

- use the benchmark percents of 1%, 10%, 25%, and 50%

- construct and use a ratio table to find what percent is equivalent to a given fraction or ratio, or vice versa

- construct and use a visual model of percent

- recognize the relationship between a fraction, a ratio, and a percent

- understand the relative nature of percent

- understand that fractions, ratios, and percents are used as comparison tools

- understand that percents are a means of standardizing to make comparisons, in static situations as well as in simple growth situations

- determine what comparison tool is most appropriate to make a comparison in a given situation: fractions, ratios, or percents

- determine which strategy for finding percent is most appropriate in a given situation

- determine whether or not percents are used appropriately in a decision-making situation

Assessment Opportunities

Keep It Clean
Decorating
On Loan

Keep It Clean
On Loan
Parking Lots

On Loan

Keep It Clean

Keep It Clean

The Best Buy
Jammin'
Decorating

Jammin'
Parking Lot

Jammin'
Parking Lot

Parking Lot

Jammin'
Parking Lot
Now It Is Your Turn

The Best Buy

Pacing

• All seven assessment problems will take approximately two 45-minute class sessions. You might have students complete the last problem, Now It Is Your Turn, as a homework assignment.

About the Mathematics

These end-of-unit assessments assess the majority of the goals of the Per Sense unit. The assessments do not ask students to use any specific strategy. Students are free to use any strategy they feel comfortable with to solve each problem.

Materials

• Assessments, pages 80, 82, 84, 86, 88, 90, and 92 of this Teacher Guide (one of each per student)
• calculators (one per student)

Planning Assessment

Students should work on these assessment problems individually if you want to evaluate each student's understanding and abilities. Make sure that you allow enough time for students to complete the problems. Students are free to solve these problems in their own way. Calculators may be used if the students so choose.

Scoring

The emphasis in scoring should be on the strategies used to solve the problems rather than on students' final answers. Since several strategies can be employed to answer many of the questions, the strategies students choose may indicate how well they understand percent concepts. For example, a student who uses a concrete strategy supported by drawings may have a deeper understanding of percents than a student who uses an abstract computation. Consider how well the students' strategies address the problem as well as how successful the students are at applying their strategies in the problem-solving process.

KEEP IT CLEAN

Use additional paper as needed.

This vacuum cleaner has a dust meter. It shows how much dust is in the dust bag. When the meter indicates that the dust bag is full, you have to change the dust bag.

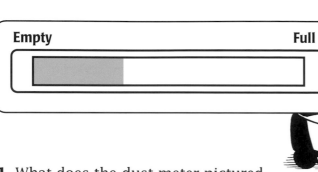

Empty   Full

1. What does the dust meter pictured above tell you? About what percent of the bag is taken up by dust? Explain how you got your answer.

2. Two weeks later dust takes up four-fifths of the bag. What will the dust meter look like? Draw it in.

Empty   Full

3. What percent of the bag is full of dust? Explain how you got your answer.

Empty   Full

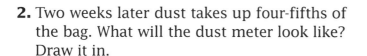

1. Accept an estimate of about 30 percent or an exact answer of $33\frac{1}{3}$ percent.

2.

Empty   Full

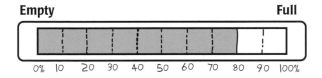

3. Eighty percent. Strategies will vary. Some students may divide the dust meter into 10 equal sections as shown below. Other students may try to find the percent equivalent for $\frac{4}{5}$.

Sample strategy:

Empty   Full

| | | | | | | | | | |
0% 10 20 30 40 50 60 70 80 90 100%

Overview Students read the dust meter of a vacuum cleaner (a fraction bar) to estimate the fraction and percent of the bag that is full.

About the Mathematics Students should be able to determine the equivalent percents for easy fractions by calculating an exact percent or making a percent estimate.

Planning Have students work individually on this assessment.

Comments about the Problems

1. This problem introduces the dust meter as a tool for estimating percents. Note whether or not students put a percent scale on the dust meter to find the percent.

2. Students are asked to make a fraction model for four-fifths. The fraction is more difficult than the one in problem **1**.

3. Here students are asked to find what percent of the dust bag is full. Notice the strategy used to find the percent. Some students may use the dust meter as a visual model (percent bar), while others may calculate the percent using benchmark fractions and percents ($\frac{1}{5}$ = 20 percent, so $\frac{4}{5}$ = 80 percent full).

THE BEST BUY

Use additional paper as needed.

1. In which of the two shops do you think the sale price of the tennis shoes is lower? Explain why you think so.

2. Is it also possible for the sale price of the shoes in the other shop to be lower? Explain your answer.

Solutions and Samples
of student work

1. Answers will vary. Students who choose Barbara's Bargain Basement must explain that Barbara's store is offering the lower sale price if the original price for the tennis shoes was the same, lower, or even a little higher than the original price at Dennis's Discount Dynasty. Students who choose Dennis's Discount Dynasty must explain that Dennis's store is offering the lower sale price if the original price of the tennis shoes at his store was already much lower than the original price at Barbara's store.

   A student might incorrectly answer that 25 percent off results in a lower price because 25 percent is less than 40 percent. Likewise, if a student responds that a 40 percent discount produces the lower sale price without mentioning the relative sizes of the two original prices, the student's response is essentially incorrect.

2. Yes, it is possible. As detailed above, it is possible that Barbara's price is the lower and equally possible that Dennis's price is lower.

Overview Students compare the percent of discount being offered for the same tennis shoes at two shoe stores to determine which store is offering the lower price.

About the Mathematics Students should understand that percents are relative numbers, and that one cannot compare them without taking into account the base to which each percent refers. They should also understand that the percent changes if the point of reference changes.

Planning Have students work individually on this assessment.

Comments about the Problems

1. Students can show that they understand that percents are relative numbers.

2. The same reasoning used in problem **1** can be used to justify the answer to this problem. The 25 percent discount offered by Dennis may actually result in a lower price if Dennis's retail price for the tennis shoes is lower than Barbara's. Therefore, this question is a "safety net question," offering students who did not adequately explain their answer in problem **1** an extra chance to show their understanding of the relative nature of percent.

JAMMIN'

Use additional paper as needed.

Jams can differ in quality. The quality often depends on the percent of fruit that is used for making the jam. The higher the percent, the higher the quality.

1. What can you say about the quality of these three cherry jams?

2. This black currant jam is sold in large and small jars. Someone forgot to put the percent of fruit on the smaller jar. Fill in this missing information. Explain your strategy for finding the percent.

3. How many grams of fruit does this jar contain? Explain how you got your answer.

1. Answers will vary. Some students may say that the cherry jam with 55 percent fruit has the highest quality because it has the most fruit.

2. The black currant jam in both jars is the same. Therefore, the percent of black currant fruit (60 percent) is also the same.

3. 270 grams. Strategies will vary. Sample student strategies:

   Strategy 1
   The fraction equivalent for 60 percent is $\frac{3}{5}$;
   $\frac{3}{5} \times 450 = 270$ grams.

   Strategy 2
   Make a percent bar and use benchmark percents to estimate 60 percent of 450.

   Strategy 3
   Using benchmark percents:
   10 percent of $450 = \frac{1}{10} \times 450 = 45$;
   $45 \times 6 = 270$ grams

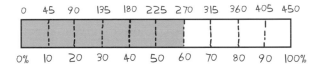

Overview Students compare the percents of fruit in three different cherry jams and find the amount of fruit in a jar of black currant jam when the percent of fruit and the total grams of jam are known.

About the Mathematics Students should understand that percents are relative numbers, and that two percents are equal if the ratios are equal, and vice versa (as in problems **2** and **3**). They should also be able to determine, by estimating or calculating, the fraction of the whole when the percent is given.

Planning Have students work individually on this assessment.

Comments about the Problems

1. This question concerns the concept of percent. Students should understand that the quality of the jams depends on the *percent* of fruit in each jar and not on the different *amounts* of fruit. A better quality jam contains a higher fraction or percent of fruit.

2. Some students may incorrectly put 40 percent on the smaller jar due to the difference in size between the two jars.

3. Note which percent strategies students use to find the answer. Some may use a percent bar to estimate; others may use the 10 percent benchmark and multiply that amount by six. Still others may see that 60 percent $= \frac{3}{5}$ and multiply to find $\frac{3}{5}$ of 450.

DECORATING THE HOUSE

Use additional paper as needed.

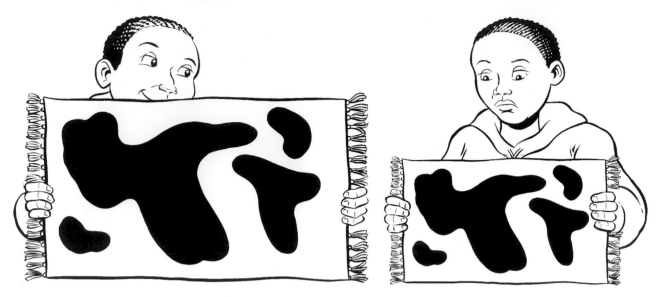

These are very special carpets. The black part is made of wool and the rest
is made of sisal. Sisal is a rope made from the fiber of sisal plant leaves. The
large carpet sold so well that the manufacturer decided to make a duplicate
in a smaller size for use as a welcome mat.

1. What percent of the large carpet is made of wool? Explain how you got
your answer.

2. Estimate the percent of the small carpet that is made of wool. Explain how
you got your answer.

1. Accept any reasonable estimate. Approximately 40 percent of the large carpet is made of wool. Strategies will also vary. Some students may draw a grid on the carpet and count how many grid squares are black. They can then use a ratio table or percent bar to estimate what percent of the total squares the black squares represent. Other students may simply make a visual estimate.

2. The percent answer here should be the same as the answer to problem **1** since the black sections of the small carpet are proportionate to those of the large carpet.

Overview Students compare two carpets to estimate what percent of each carpet is made of wool.

About the Mathematics This assessment is similar to *Jammin'*. Again, the concept that two equal ratios represent the same percent is emphasized.

Planning Have students work individually on this assessment.

Comments about the Problems

1. Students will use different estimation strategies. Some students may visually estimate the percent of wool while others may draw a grid on the carpet and count the number of black squares to find a more precise percent estimate.

2. Students should understand that both carpets have the same percent of wool since the ratio of wool to the entire rug in both carpets is the same.

ON LOAN

Use additional paper as needed.

1. Complete the Library Information sign. An estimation of the percent will do. Explain how you got your answer.

LIBRARY INFORMATION

TOTAL BOOKS: 6 9 9 7

ON LOAN: 2 8 1 3

% ON LOAN

1. Accept any reasonable estimate close to 40 percent. Strategies will vary.

Sample strategies:

Strategy 1
Construct a ratio table after rounding both numbers.
Books on Loan: $2,813 \approx 2,800$
Total Books: $6,997 \approx 7,000$
Forty out of 100 is about 40 percent.

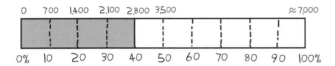

Strategy 2
Construct a percent bar and use benchmark percents after rounding both numbers.
Books on Loan: $2,813 \approx 2,800$
Total Books: $6,997 \approx 7,000$

Strategy 3
After rounding both numbers, express the ratio of the number of books on loan to the total number of books as a fraction. Then reduce this fraction to a fraction with a denominator of 100.

Books on Loan: $2,813 \approx 2,800$
Total Books: $6,997 \approx 7,000$

$$\frac{2,800}{7,000} = \frac{280}{700} = \frac{40}{100} = 40 \text{ percent}$$

Overview Students estimate the percent of library books on loan.

About the Mathematics Students must find the percent for large numbers. They should be able to determine the percent, by estimating or calculating, since the part and the whole are given.

Planning Have students work individually on this assessment.

Comments about the Problems

1. Some students may round the numbers and estimate the percent using fractions. Others may make a percent estimate using a percent bar. Still other students may calculate the exact percent and then round their answers.

PARKING LOTS

Use additional paper as needed.

Some parking lots have signs
that tell how occupied the lots
are. When 90% of the spaces are
occupied, a red light will go on.
Below are signs for two of
these lots.

PARKING LOT A

NUMBER OF SPACES	200
OCCUPIED	183

90% LIGHT

PARKING LOT B

NUMBER OF SPACES	300
OCCUPIED	255

90% LIGHT

1. Which parking lot is more occupied? Explain how you got your answer.

2. For each parking lot, figure out whether the red light is on or off.
Explain how you got your answer.

Solutions and Samples
of student work

1. Answers will vary. Students who compare only the number of empty spaces in both lots might say that Lot A is more occupied because it has only 17 empty spaces, while Lot B has 45 empty spaces. Students who compare the ratios of occupied spaces to total spaces in both lots will say that Lot A is more occupied because it is about 92 percent full, while Lot B is only about 85 percent full. Accept percent estimates as well as exact percent answers.

   Strategies will vary. Students may use any of the percent strategies presented in the unit: ratio tables, percent bars, benchmark percents, or the relationships between fractions, ratios, and percents.

2. The red light for Lot A is on. The red light for Lot B is off. Students' explanations will vary, depending on the strategy they used in problem **1.** If students calculated correct percent estimates or exact percents in problem 1, they can use these answers to justify their answer for problem **2.**

Hints and Comments

Overview Students compare the number of occupied spaces out of the total spaces in two parking lots to determine which lot is more occupied.

Planning Have students work individually on this assessment.

Comments about the Problems

1. This question assesses students' ability to compare different "parts" of different "wholes." Several strategies are possible. They can use ratios, fractions, or percents to find the answer. Some students may make an absolute comparison just by looking at the numbers. They may say that Lot B is more occupied, showing that they do not understand that the percents of occupied spaces in the lots is a relative concept.

2. Students can use any percent strategy. Some students will estimate that 183 out of 200 is a little bit more than 90 percent and that 255 our of 300 is about five out of six, or between 80 and 85 percent. Others may find 90 percent of 200 and 90 percent of 300 and compare these results with the numbers on the two signs.

NOW IT IS YOUR TURN

Use additional paper as needed.

1. Think up an easy percent problem and a difficult one. Write each problem and show strategies and solutions for both. Explain why you think your first problem is easy and your second problem is more difficult.

This is my easy percent problem.

I think this is an easy problem because ...

This is my difficult percent problem.

I think this is a difficult problem because ...

1. Problems and solutions will vary.

Overview Students write their own percent problems—one easy problem and one difficult problem. They also explain what makes their easy problem simple to solve and what makes their difficult problem hard to solve.

Planning Have students work individually on this assessment.

Comments about the Problems

1. This assessment problem will tell you what types of problems your students consider easy or difficult. Students may write their own original percent problem or may recall problems from this unit that they found easy or difficult.

Per Sense
Glossary

The Glossary defines all vocabulary words listed on the Section Opener pages. It includes the mathematical terms that may be new to students, as well as words having to do with the contexts introduced in the unit. (Note: The student book has no glossary in order to allow students to construct their own definitions, based on their personal experiences with the unit activities.)

The definitions below are specific to the use of the terms in this unit. The page numbers given are from this Teacher Guide.

benchmark percent (p. 60) a percent that is easily found, such as 10%, 25%, and 50%

chance (p. 6) the likelihood that a given event will happen

discount (p. 8) a reduction in price

graph (p. 54) a visual display of information

half (p. 8) one part of two equal parts; 50%

increase (p. 12) to become greater

percent (p. 6) one part of one hundred equal parts

percent bar (p. 38) a bar that shows a range from zero to a certain quantity along the top and from 0% to 100% along the bottom. This bar helps in relating parts to percents.

poll (p. 58) to take a survey

ratio table (p. 28) a table in which the numbers in each column have the same ratio

reduce (p. 18) to make smaller

survey (p. 58) the act of asking questions in order to collect data

Blackline Masters

Dear Family,

Very soon your child will begin the *Mathematics in Context* unit called *Per Sense.* Below is a letter to your child that opens the unit, describing the unit and its goals.

You can help your child relate the class work to his or her own life by talking about percents that you encounter together in stores, weather forecasts, banks, and other places.

Scan your local newspaper for situations involving percents. Discuss the meaning of each situation with your child and the information given by the percent.

Help your child see percents as parts of a whole. Help him or her discover the relationship between percents and fractions.

Let your child show you how he or she can estimate and find percents using tables, percent bars, and "benchmark" (common) percents in combination (for instance, to find 60%, you can find 50% and 10% and add them together).

Sincerely,

The Mathematics in Context Development Team

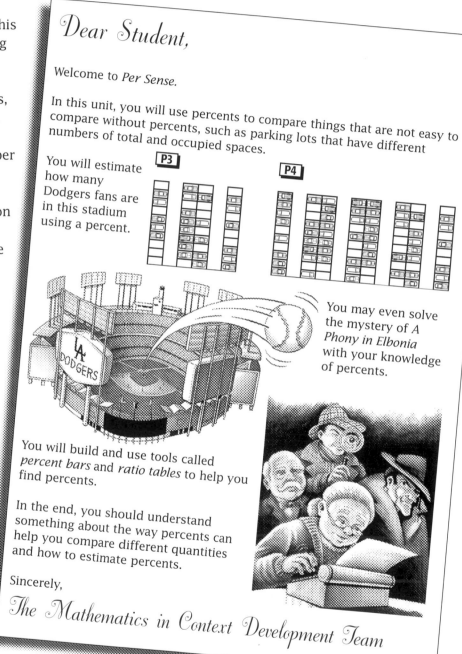

Dear Student,

Welcome to *Per Sense.*

In this unit, you will use percents to compare things that are not easy to compare without percents, such as parking lots that have different numbers of total and occupied spaces.

You will estimate how many Dodgers fans are in this stadium using a percent.

You may even solve the mystery of *A Phony in Elbonia* with your knowledge of percents.

You will build and use tools called *percent bars* and *ratio tables* to help you find percents.

In the end, you should understand something about the way percents can help you compare different quantities and how to estimate percents.

Sincerely,

The Mathematics in Context Development Team

	Audience

11. Shade in the audience section of the theater to show what percent of the theater you think would be filled for each show.

a. A pop concert

	Audience

b. A historical play

	Audience

c. A fashion show

Event	Estimated Number of Visitors	How I Found This Answer
The Pop Concert		
The Historical Play		
The Fashion Show		
The Choral Concert		
The Jazz Concert		
The Rap Concert		

Name_____

Use with *Per Sense,* page 8.

16. Make drawings to express the following percents:

    a. 50 percent of the students are girls

    b. 25 percent of the flowers are red

    c. 100 percent of the cookies are broken

17. Complete the drawing.

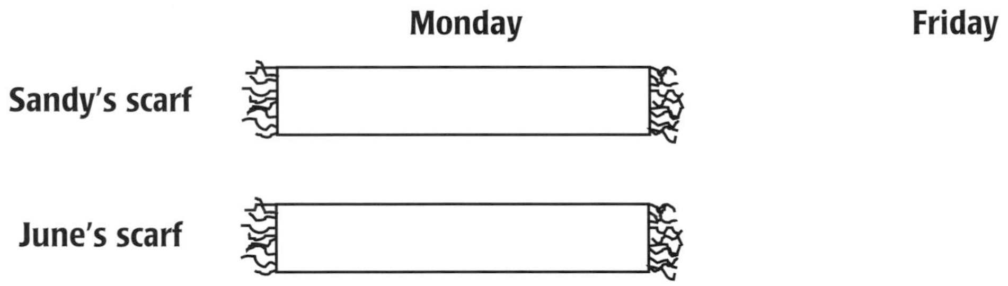

Monday   **Friday**

Sandy's scarf

June's scarf

KEY

☐ Available
🚗 Occupied

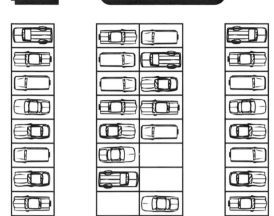

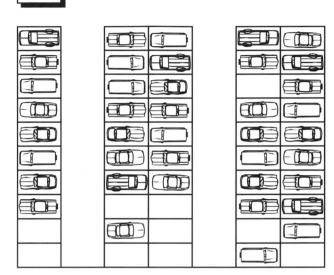

5. Fill in the parking lot signs.

P1

Number of Spaces: ____

Number Occupied: ____

Number Available: ____

Fraction Occupied: ____

Fraction Available: ____

P2

Number of Spaces: ____

Number Occupied: ____

Number Available: ____

Fraction Occupied: ____

Fraction Available: ____

6. Shade in each box to show what fraction of each lot is occupied.

P1

P2

Use with *Per Sense,* pages 11 and 13.

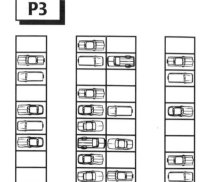

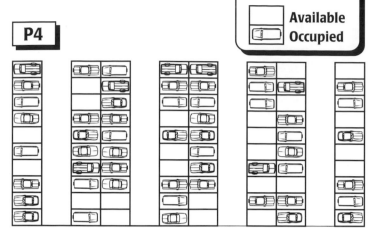

KEY

☐ Available
🚗 Occupied

8. Complete the parking lot signs for P3 and P4.

P3

Number of Spaces: _____

Number Occupied: _____

Number Available: _____

Fraction Occupied: _____

Fraction Available: _____

P4

Number of Spaces: _____

Number Occupied: _____

Number Available: _____

Fraction Occupied: _____

Fraction Available: _____

9. Shade each bar to show the number of cars in each lot.

0 cars   40 cars

P3 [   ]

0 cars   80 cars

P4 [   ]

10. Complete the tables below.

	P3	Lot A	Lot B	Lot C
Spaces Occupied	24		36	
Total Spaces	40	20		10

	P4	Lot D	Lot E	Lot F
Spaces Occupied	56	28		35
Total Spaces	80		10	

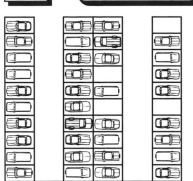
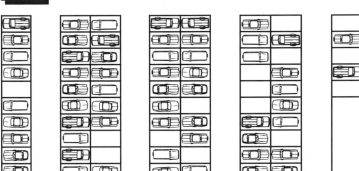

KEY

☐ Available
🚗 Occupied

P5

P6

12. Fill in the parking lot signs.

P5

Number of Spaces: ____

Number Occupied: ____

Number Available: ____

Fraction Occupied: ____

Fraction Available: ____

P6

Number of Spaces: ____

Number Occupied: ____

Number Available: ____

Fraction Occupied: ____

Fraction Available: ____

13. Shade each bar to show what fraction of each lot is occupied.

0 cars   40 cars

P5 [   ]

0 cars   75 cars

P6 [   ]

14. Complete the tables below.

	P5	Lot G	Lot H	Lot I
Spaces Occupied	36		9	
Total Spaces	40	20		60

	P6	Lot J	Lot K	Lot L
Spaces Occupied	60			
Total Spaces	75	25	50	100

Name_____

Use with *Per Sense*, page 14.

19. Shade the parking lot bars to show what percent of the parking lot is occupied during each two–hour period.

Time	Total Number of Occupied Spaces	Percent Bar
8 A.M. – 10 A.M.	90	0% no cars 100% full
10 A.M. – 12 P.M.	120	
12 P.M. – 2 P.M.	180	
2 P.M. – 4 P.M.	240	
4 P.M. – 6 P.M.	300	
6 P.M. – 8 P.M.	320	
8 P.M. – 10 P.M.	400	
10 P.M. – 12 A.M.	400	
12 A.M. – 2 A.M.	100	
2 A.M. – 4 A.M.	50	
4 A.M. – 6 A.M.	40	
6 A.M. – 8 A.M.	60	

office hours

Name_____

20. What fraction of each lot or floor is occupied?
Express each fraction as a percent. Then
complete the parking lot signs.

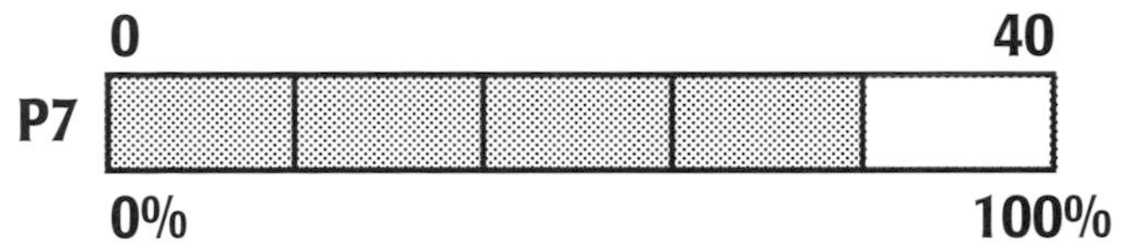

0   40

P7 [   ]

0%   100%

P7
Number of Spaces: ——
Number Occupied: ____
Number Available: ____
Fraction Occupied: ____
Fraction Available: ____
Percent Occupied: ____
Percent Available: ____

0   125

P8 [   ]

0%   100%

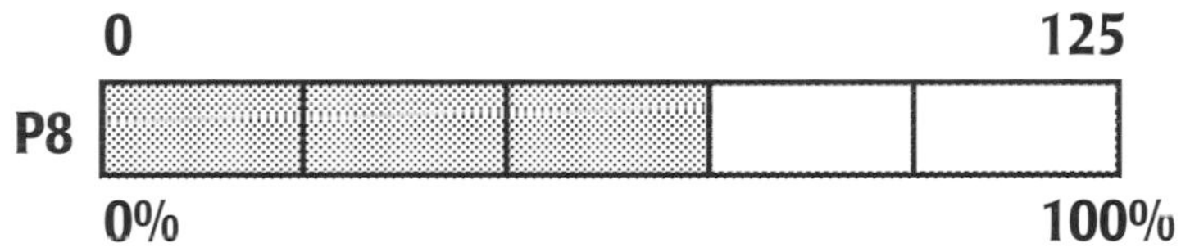

P8
Number of Spaces: ——
Number Occupied: ____
Number Available: ____
Fraction Occupied: ____
Fraction Available: ____
Percent Occupied: ____
Percent Available: ____

P9
0   80

[   ]

0%   100%

[   ]

[   ]

[   ]

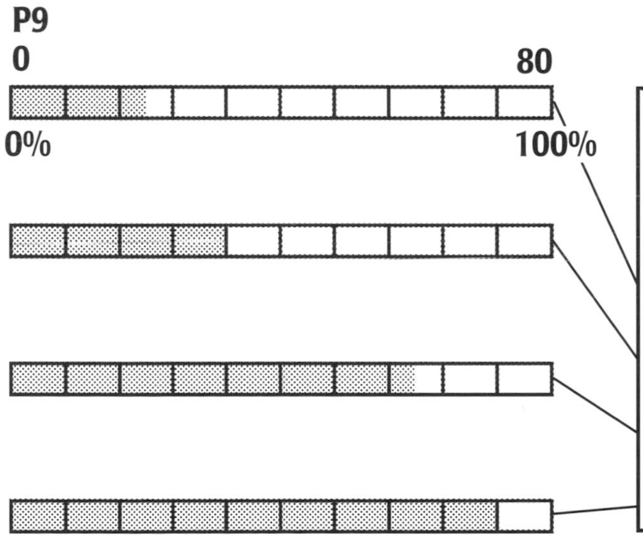

| P9 | Number of Spaces on Each Floor: 80 | | |
Occupied	Available	Fraction Occupied	Percent Occupied
____	____	____	____
____	____	____	____
____	____	____	____
____	____	____	____

Name_____

Use with *Per Sense,* page 15.

21. Below are three signs for each parking lot. Fill them in with different numbers of occupied and available spaces so that for each parking lot, all three signs show the same *percent* of occupied spaces.

P11	P11	P11
Number of Spaces: _____	Number of Spaces: _____	Number of Spaces: _____
Number Occupied: _____	Number Occupied: _____	Number Occupied: _____
Number Available: _____	Number Available: _____	Number Available: _____
Percent Occupied: _50_	Percent Occupied: _50_	Percent Occupied: _50_
Percent Available: _____	Percent Available: _____	Percent Available: _____

P12	P12	P12
Number of Spaces: _____	Number of Spaces: _____	Number of Spaces: _____
Number Occupied: _____	Number Occupied: _____	Number Occupied: _____
Number Available: _____	Number Available: _____	Number Available: _____
Percent Occupied: _25_	Percent Occupied: _25_	Percent Occupied: _25_
Percent Available: _____	Percent Available: _____	Percent Available: _____

P13	P13	P13
Number of Spaces: _____	Number of Spaces: _____	Number of Spaces: _____
Number Occupied: _____	Number Occupied: _____	Number Occupied: _____
Number Available: _____	Number Available: _____	Number Available: _____
Percent Occupied: _10_	Percent Occupied: _10_	Percent Occupied: _10_
Percent Available: _____	Percent Available: _____	Percent Available: _____

20. Look at the two graphs on page 23. Assume that Group A is the Dodgers. Fill in the table with the appropriate number of fans for each category.

	Total Number of Spectators	Number of Dodgers Fans	Number of Giants Fans
Half an Hour Before the Game Started			
During the Game	60,000	55,200	4,800
Fifteen Minutes After the Game Ended			
One Half Hour After the Game Ended			

Explain how you got your answers. Use additional paper as needed.

Use with *Per Sense,* pages 23 and 24.

21. Complete the diagram. Fill in the boxes for the percent and the number of Giants fans in the stadium 15, 20, and 27 minutes after the game.

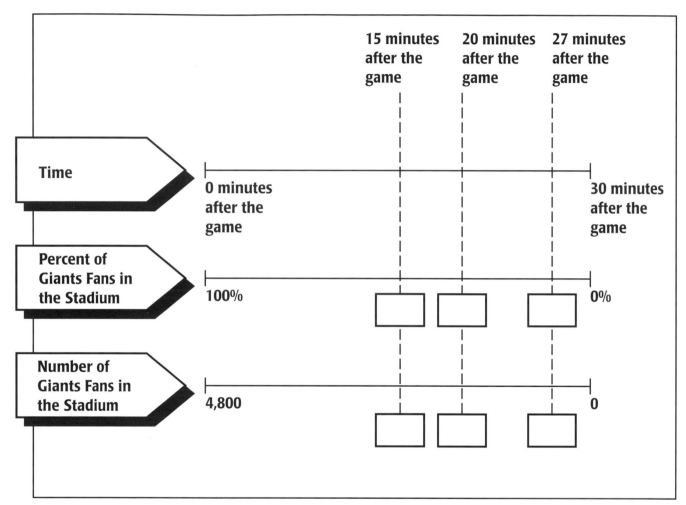

23. a. Fill in the percent bars. What is the percent of students for each grade level who are interested in going to Noah's Ark?

Preliminary Poll

Grade 6	
Interviewed Students	40
Interested	26 is ___ %

0   40

0%   100%

Grade 7	
Interviewed Students	35
Interested	24 is ___ %

0   35

0%   100%

Grade 8	
Interviewed Students	30
Interested	14 is ___ %

0   30

0%   100%

24. a. Fill in the percent bars. What is the percent of students in each grade level who favor going to Noah's Ark now?

Entire Student Body

Grade 6	
Interviewed Students	1,045
Interested	689 is ___ %

0   1,045

0%   100%

Grade 7	
Interviewed Students	1,839
Interested	1,361 is ___ %

0   1,839

0%   100%

Grade 8	
Interviewed Students	1,495
Interested	912 is ___ %

0   1,495

0%   100%

Use with *Per Sense,* page 26.

27. Find the percent of dropouts from the marathon portion of the triathlon. Describe your strategies.

Year of Marathon	Number of Competitors	Number of Dropouts	Percent of Dropouts	Describe Your Strategy
1988	1,340	670		
1989	1,621	392		
1990	1,793	180		
1991	1,603	91		
1992	1,400	350		

Name _____ **Date** _____

Use additional paper as needed.

This vacuum cleaner has a dust meter. It shows how much dust is in the dust bag. When the meter indicates that the dust bag is full, you have to change the dust bag.

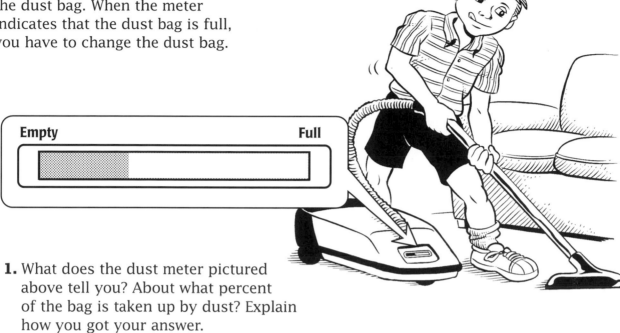

Empty   **Full**

1. What does the dust meter pictured above tell you? About what percent of the bag is taken up by dust? Explain how you got your answer.

2. Two weeks later dust takes up four-fifths of the bag. What will the dust meter look like? Draw it in.

Empty   **Full**

3. What percent of the bag is full of dust? Explain how you got your answer.

Empty   **Full**

Name _____   Date _____

THE BEST BUY

Use additional paper as needed.

1. In which of the two shops do you think the sale price of the tennis shoes is lower? Explain why you think so.

2. Is it also possible for the sale price of the shoes in the other shop to be lower? Explain your answer.

Use additional paper as needed.

Jams can differ in quality. The quality often depends on the percent of fruit that is used for making the jam. The higher the percent, the higher the quality.

Cherry Jam 45% Fruit

Cherry Jam 43% fruit

Cherry Jam 55% Fruit

1. What can you say about the quality of these three cherry jams?

2. This black currant jam is sold in large and small jars. Someone forgot to put the percent of fruit on the smaller jar. Fill in this missing information. Explain your strategy for finding the percent.

Black Currant Jam 225g % fruit

Black Currant Jam 450g 60% fruit

3. How many grams of fruit does this jar contain? Explain how you got your answer.

Black Currant Jam 450g 60% fruit

DECORATING THE HOUSE

Use additional paper as needed.

These are very special carpets. The black part is made of wool and the rest is made of sisal. Sisal is a rope made from the fiber of sisal plant leaves. The large carpet sold so well that the manufacturer decided to make a duplicate in a smaller size for use as a welcome mat.

1. What percent of the large carpet is made of wool? Explain how you got your answer.

2. Estimate the percent of the small carpet that is made of wool. Explain how you got your answer.

Use additional paper as needed.

1. Complete the Library Information sign. An estimation of the percent will do. Explain how you got your answer.

LIBRARY

LIBRARY INFORMATION
TOTAL BOOKS: 6 9 9 7
ON LOAN: 2 8 1 3

% ON LOAN

PARKING LOTS

Use additional paper as needed.

Some parking lots have signs that tell how occupied the lots are. When 90% of the spaces are occupied, a red light will go on. Below are signs for two of these lots.

PARKING LOT A

| NUMBER OF SPACES | 200 |
| OCCUPIED | 183 |

90% LIGHT

PARKING LOT B

| NUMBER OF SPACES | 300 |
| OCCUPIED | 255 |

90% LIGHT

1. Which parking lot is more occupied? Explain how you got your answer.

2. For each parking lot, figure out whether the red light is on or off.
Explain how you got your answer.

NOW IT IS YOUR TURN

Use additional paper as needed.

1. Think up an easy percent problem and a difficult one. Write each problem and show strategies and solutions for both. Explain why you think your first problem is easy and your second problem is more difficult.

This is my easy percent problem.

I think this is an easy problem because ...

This is my difficult percent problem.

I think this is a difficult problem because ...

Section A

1. a–c. Answers will vary. Students may write sentences that represent percents but do not actually include percents. For example, a sentence for 50% might be: "Only half of the students are in class today."

2. Drawings will vary, but should represent 100%, 50%, and 25%.

3. Definitions will vary. See glossary on page 94 of this Teacher Guide.

Section B

1. Yes, the statement is true. Both parking lots are 75% occupied.

2.

Parking Lot X		**Parking Lot Y**	
Number of Spaces:	16	Number of Spaces:	40
Number Occupied:	12	Number Occupied:	30
Number Available:	4	Number Available:	10
Fraction Occupied:	$\frac{12}{16}$ or $\frac{3}{4}$	Fraction Occupied:	$\frac{30}{40}$ or $\frac{3}{4}$
Fraction Available:	$\frac{4}{16}$ or $\frac{1}{4}$	Fraction Available:	$\frac{10}{40}$ or $\frac{1}{4}$

3.

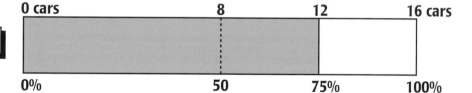

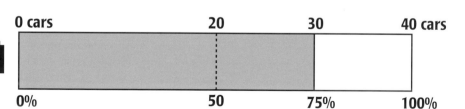

4. Answers will vary. The ratio of occupied spaces to total spaces should be 3 to 4. Sample solution:

	Lot X	Lot A	Lot B	Lot C	Lot D	Lot E	Lot F	Lot G	Lot H
Spaces Occupied	12	60	9	3	75	6	15	30	300
Total Spaces	16	80	12	4	100	8	20	40	400

5. Answers will vary. The ratio of occupied spaces to total spaces should be 2 to 3. Sample solution:

	Lot J	Lot K	Lot L	Lot M	Lot N	Lot P	Lot Q	Lot R	Lot S
Spaces Occupied	16	2	6	8	20	200	10	40	80
Total Spaces	24	3	9	12	30	300	15	60	120

6. Answers will vary. Sample responses may include relative comparisons as well as actual comparisons. For example:

Relative Comparison

Parking Lot X is $\frac{3}{4}$ occupied. Parking Lot J is $\frac{2}{3}$ occupied. Since $\frac{3}{4}$ is larger than $\frac{2}{3}$, Parking Lot X is more occupied than Parking Lot J.

Actual Comparison

Parking Lot X has 12 cars in it, and Parking Lot J has 16 cars in it. Since Parking Lot J has more cars, it is more occupied than Parking Lot X.

Section C

1. a. 60%; ratio tables will vary.

		÷1000	÷4	×10
Lions Fans	24,000	24	6	60
People in Stadium	40,000	40	10	100

b. 40%

2. 90%; ratio tables will vary.

		÷1000	÷4	×10
Lions Fans	36,000	36	9	90
People in Stadium	40,000	40	10	100

3. 10%; ratio tables will vary.

		÷1000	÷4	×10
Lions Fans	4,000	4	1	10
People in Stadium	40,000	40	10	100

4. a. 75%; ratio tables will vary.

		÷1000	÷8	× 25
Eagles Fans	24,000	24	3	75
People in Stadium	32,000	32	4	100

b. 25%

5. about 33%; ratio tables will vary.

		÷1000	÷12	× 11	× 3
Buttons Sold	11,651	≈12,000	12	1	11	33
People in Stadium	35,523	≈36,000	36	3	33	99

Section D

1. Answers will vary; possible estimates:

   a. $4.40

   b. $1.60

   c. $0.80

   d. $6.90

   e. $3.90

2. Explanations will vary. A common strategy is to round the bill amount to the nearest dollar. Then mentally calculate 10% of that amount.

3. Answers may vary; possible estimates:

   a. $6.60

   b. $2.40

   c. $1.20

   d. $10.35

   e. $5.85

4. Explanations will vary. Some students may look back at problem **1** and calculate 1.5 times their answers to that problem, since a 15% tip is one and one-half times larger than a 10% tip. Others may round each total to the nearest whole number. They might then take 10% of that whole number, halve the 10% to get 5%, and then add 10% and 5% to find 15%.

Cover

Design by Ralph Paquet/Encyclopædia Britannica Educational Corporation.

Collage by Koorosh Jamalpur/KJ Graphics.

Title Page

Illustration by Brent Cardillo/Encyclopædia Britannica Educational Corporation.

Illustrations

6, 8, 10, 12 (bottom) Paul Tucker/Encyclopædia Britannica Educational Corporation; **12 (top)** Phil Geib/Encyclopædia Britannica Educational Corporation; **14, 16** Jerome Gordon; **18, 20, 24, 26, 28, 30, 34, 38** Phil Geib/Encyclopædia Britannica Educational Corporation; **42, 44, 46, 48** Brent Cardillo/Encyclopædia Britannica Educational Corporation; **50, 52, 58, 64** Phil Geib/Encyclopædia Britannica Educational Corporation; **66** Paul Tucker/Encyclopædia Britannica Educational Corporation; **70** Brent Cardillo/Encyclopædia Britannica Educational Corporation; **72** Phil Geib/Encyclopædia Britannica Educational Corporation; **74** Brent Cardillo/Encyclopædia Britannica Educational Corporation; **76** Phil Geib/Encyclopædia Britannica Educational Corporation; **80, 82, 84, 86, 88, 90, 92** Brent Cardillo/Encyclopædia Britannica Educational Corporation.

Photographs

24 © Frank Cezus/Tony Stone Images; **32** © George Mars Cassidy/Tony Stone Images; **42** © Baron Wolman/Tony Stone Images; **56** © Mary Wolf/Tony Stone Images.

Special Thanks

Major League Baseball trademarks and copyrights used with permission from Major League Baseball Properties, Inc.